# The New Engineering Contract
## A Commentary

**Also from Blackwell Science**

*Liquidated Damages and Extensions of Time in Construction Contracts*
Brian Eggleston
0–632–03295–2

*The ICE Conditions: Sixth Edition*
Brian Eggleston
0–632–03092–3

*The ICE Design and Construct Contract*
Brian Eggleston
0–632–03092–5

*Model Forms of Contract for Electrical and Mechanical Plant*
Brian Eggleston
0–632–03803–9

# The New Engineering Contract

## A Commentary

Brian Eggleston
*CEng, FICE, FIStructE, FCIArb*

Blackwell
Science

© Brian Eggleston 1996
Blackwell Science Ltd
Editorial Offices:
Osney Mead, Oxford OX2 0EL
25 John Street, London WC1N 2BL
23 Ainslie Place, Edinburgh EH3 6AJ
238 Main Street, Cambridge
  Massachusetts 02142, USA
54 University Street, Carlton
  Victoria 3053, Australia

Other Editorial Offices:
Arnette Blackwell SA
  224, Boulevard Saint Germain
  75007 Paris, France

Blackwell Wissenschafts-Verlag GmbH
  Kurfürstendamm 57
  10707 Berlin, Germany

Zehetnergasse 6
A-1140 Wien
Austria

All rights reserved. No part of
this publication may be reproduced,
stored in a retrieval system, or
transmitted, in any form or by any
means, electronic, mechanical,
photocopying, recording or otherwise,
except as permitted by the UK
Copyright, Designs and Patents Act
1988, without the prior permission
of the publisher.

First published 1996

Set in 10 on 12pt Palatino
by DP Photosetting, Aylesbury, Bucks
Printed and bound in Great Britain
by Hartnolls Ltd, Bodmin, Cornwall.

The Blackwell Science logo is a
trade mark of Blackwell Science Ltd,
registered at the United Kingdom
Trade Marks Registry.

DISTRIBUTORS

Marston Book Services Ltd
PO Box 269
Abingdon
Oxon OX14 4YN
(*Orders:* Tel: 01235 465500
         Fax: 01235 465555)

USA
Blackwell Science, Inc.
238 Main Street
Cambridge, MA 02142
(*Orders:* Tel: 800 215-1000
              617 876-7000
         Fax: 617 492-5263)

Canada
Copp Clark, Ltd
2775 Matheson Blvd East
Mississauga, Ontario
Canada, L4W 4P7
(*Orders*: Tel: 800 263-4374
              905 238-6074)

Australia
Blackwell Science Pty Ltd
54 University Street
Carlton, Victoria 3053
(*Orders:* Tel: 03 9347 0300
         Fax: 03 9349 3016)

A catalogue record for this title
is available from the British Library

ISBN 0–632–04065–3

Library of Congress
Cataloging-in-publication Data
Eggleston, Brian, CEng.
  The new engineering contract: a
  commentary/Brian Eggleston.
     p.   cm.
  Includes index.
  ISBN 0–632–04065–3
  1. Engineering contracts—Great Britain.
I. Title.
KD1641.E354  1996
343.41'07862—dc20
[344.1037862]                96-14603
                                 CIP

# Contents

Preface ... xi

1 **Introduction** ... 1
    1.1 Background ... 1
    1.2 Aims and objectives ... 2
    1.3 Structure of the NEC ... 4
    1.4 Special characteristics of the NEC ... 6

2 **The main options** ... 8
    2.1 Introduction ... 8
    2.2 Contract strategy ... 9
    2.3 Option A – priced contract with activity schedule ... 16
    2.4 Option B – priced contract with bill of quantities ... 17
    2.5 Target contracts generally ... 18
    2.6 Options C and D – target contracts ... 21
    2.7 Option E – cost reimbursable contract ... 22
    2.8 Option F – management contract ... 23

3 **The secondary options** ... 25
    3.1 Introduction ... 25
    3.2 Option G – performance bond ... 27
    3.3 Option H – parent company guarantee ... 29
    3.4 Option J – advanced payment to the contractor ... 30
    3.5 Option K – multiple currencies ... 31
    3.6 Option L – sectional completion ... 31
    3.7 Option M – limitation of the contractor's liability for his design to reasonable skill and care ... 32
    3.8 Option N – price adjustment for inflation ... 34
    3.9 Option P – retention ... 34
    3.10 Option Q – bonus for early completion ... 35
    3.11 Option R – delay damages ... 37
    3.12 Option S – low performance damages ... 41
    3.13 Option T – changes in the law ... 43
    3.14 Option U – The Construction (Design and Management) Regulations 1994 ... 44
    3.15 Option V – trust fund ... 45

## Contents

**4 The contract documents** — 47
- 4.1 Introduction — 47
- 4.2 Sequence of documentation — 48
- 4.3 Essential contract documents — 49
- 4.4 Identified and defined terms — 51
- 4.5 The contract date — 51
- 4.6 Works information — 53
- 4.7 Site information — 56
- 4.8 Contract data — 58
- 4.9 Schedules of cost components — 58
- 4.10 Ambiguities and inconsistencies in the contract documents — 62
- 4.11 Schedule of clauses referring to the works information — 62

**5 The key players** — 67
- 5.1 Introduction — 67
- 5.2 Others — 68
- 5.3 Actions — 70
- 5.4 The employer — 73
- 5.5 Express obligations of the employer — 74
- 5.6 The project manager — 75
- 5.7 Express duties of the project manager — 77
- 5.8 The supervisor — 82
- 5.9 Express duties of the supervisor — 82
- 5.10 Communications — 83
- 5.11 The project manager and the supervisor — 87

**6 General core clauses** — 89
- 6.1 Introduction — 89
- 6.2 Actions — 89
- 6.3 Identified and defined terms — 91
- 6.4 Interpretation and the law — 98
- 6.5 Communications — 99
- 6.6 The project manager and the supervisor — 99
- 6.7 Adding to the working areas — 99
- 6.8 Early warning — 100
- 6.9 Ambiguities and inconsistencies — 103
- 6.10 Health and safety — 103
- 6.11 Illegal and impossible requirements — 104

**7 Obligations and responsibilities of the contractor** — 107
- 7.1 Introduction — 107
- 7.2 Design obligations, responsibilities and liabilities — 110
- 7.3 Providing the works — 113

|  |  |  |
|---|---|---|
| 7.4 | The contractor's design | 115 |
| 7.5 | Using the contractor's design | 118 |
| 7.6 | Design of equipment | 119 |
| 7.7 | People | 119 |
| 7.8 | Co-operation | 121 |
| 7.9 | Subcontracting | 122 |
| 7.10 | Approval from others | 125 |
| 7.11 | Access to the work | 126 |
| 7.12 | Instructions | 126 |
| 7.13 | Express obligations of the contractor | 127 |
| 7.14 | Express prohibitions on the contractor | 131 |

## 8 Time (and related matters) — 132

| | | |
|---|---|---|
| 8.1 | Introduction | 132 |
| 8.2 | Starting and completion | 134 |
| 8.3 | The programme | 136 |
| 8.4 | Revising the programme | 140 |
| 8.5 | Shortened programmes | 142 |
| 8.6 | Possession of the site | 143 |
| 8.7 | Instructions to stop or not to start work | 146 |
| 8.8 | Take over | 148 |
| 8.9 | Acceleration | 151 |

## 9 Testing and defects — 153

| | | |
|---|---|---|
| 9.1 | Introduction | 153 |
| 9.2 | Definitions and certificates | 157 |
| 9.3 | Tests and inspections | 160 |
| 9.4 | Testing and inspection before delivery | 163 |
| 9.5 | Searching and notifying defects | 164 |
| 9.6 | Correcting defects | 166 |
| 9.7 | Accepting defects | 168 |
| 9.8 | Uncorrected defects | 169 |

## 10 Payments — 171

| | | |
|---|---|---|
| 10.1 | Introduction | 171 |
| 10.2 | Assessing the amount due | 174 |
| 10.3 | Payments | 177 |
| 10.4 | Actual cost | 180 |
| 10.5 | Payments – main option A | 181 |
| 10.6 | Payments – main option B | 184 |
| 10.7 | Payments – main option C | 186 |
| 10.8 | Payments – main option D | 190 |
| 10.9 | Payments – main option E | 191 |
| 10.10 | Payments – main option F | 192 |

## Contents

| | | | |
|---|---|---|---|
| **11** | **Compensation events** | | **194** |
| | 11.1 | Introduction | 194 |
| | 11.2 | The compensation events | 199 |
| | 11.3 | Notifying compensation events | 210 |
| | 11.4 | Quotations for compensation events | 216 |
| | 11.5 | Assessing compensation events | 219 |
| | 11.6 | The project manager's assessments | 225 |
| | 11.7 | Implementing compensation events | 227 |
| **12** | **Title** | | **230** |
| | 12.1 | Introduction | 230 |
| | 12.2 | Employer's title to equipment, plant and materials | 231 |
| | 12.3 | Marking equipment, plant and materials | 232 |
| | 12.4 | Removing equipment | 233 |
| | 12.5 | Objects and materials within the site | 233 |
| **13** | **Risks and insurances** | | **235** |
| | 13.1 | Introduction | 235 |
| | 13.2 | The employer's risks | 240 |
| | 13.3 | The contractor's risks | 243 |
| | 13.4 | Repairs | 243 |
| | 13.5 | Indemnity | 244 |
| | 13.6 | Insurance cover | 244 |
| | 13.7 | Insurance policies | 245 |
| | 13.8 | Contractor's failure to insure | 246 |
| | 13.9 | Insurance by the employer | 246 |
| **14** | **Disputes** | | **248** |
| | 14.1 | Introduction | 248 |
| | 14.2 | The nature and scope of adjudication | 248 |
| | 14.3 | The procedural requirements of the NEC | 249 |
| | 14.4 | The power to review decisions | 250 |
| | 14.5 | Settlement of disputes | 251 |
| | 14.6 | The adjudication | 254 |
| | 14.7 | The adjudicator | 255 |
| | 14.8 | Review by the tribunal | 256 |
| **15** | **Termination** | | **259** |
| | 15.1 | Introduction | 259 |
| | 15.2 | Summary of the NEC termination provisions | 262 |
| | 15.3 | Termination for 'any reason' | 263 |
| | 15.4 | Termination under clause 94 of the NEC | 265 |
| | 15.5 | Reasons for termination under the NEC | 267 |
| | 15.6 | Procedures on termination under the NEC | 270 |
| | 15.7 | Amounts due on termination under the NEC | 271 |

| 16 | The NEC engineering and construction subcontract | **273** |
|---|---|---|
| | 16.1 Introduction | 273 |
| | 16.2 Core clauses – general | 275 |
| | 16.3 Core clauses – the subcontractor's main responsibilities | 276 |
| | 16.4 Core clauses – time | 276 |
| | 16.5 Core clauses – testing and defects | 277 |
| | 16.6 Core clauses – payment | 278 |
| | 16.7 Core clauses – compensation events | 278 |
| | 16.8 Core clauses – title | 279 |
| | 16.9 Core clauses – risks and insurance | 279 |
| | 16.10 Core clauses – disputes and termination | 280 |
| **17** | **The professional services contract** | **282** |
| | 17.1 Introduction | 282 |
| | 17.2 The main options | 283 |
| | 17.3 The secondary options | 285 |
| | 17.4 Core clauses – general | 287 |
| | 17.5 Core clauses – the parties' main responsibilities | 289 |
| | 17.6 Core clauses – time | 290 |
| | 17.7 Core clauses – quality | 291 |
| | 17.8 Core clauses – payment | 291 |
| | 17.9 Core clauses – compensation events | 292 |
| | 17.10 Core clauses – title | 294 |
| | 17.11 Core clauses – risks and insurance | 294 |
| | 17.12 Core clauses – disputes and termination | 295 |
| **18** | **The adjudicator's contract** | **298** |
| | 18.1 Introduction | 298 |
| | 18.2 Appointment of the adjudicator | 300 |
| | 18.3 Joinder provisions | 301 |
| | 18.4 Section 1 – general | 302 |
| | 18.5 Section 2 – adjudication | 303 |
| | 18.6 Section 3 – payment | 304 |
| | 18.7 Section 4 – title | 304 |
| | 18.8 Section 5 – risks | 305 |
| | 18.9 Section 6 – termination | 305 |
| | *Table of cases* | *307* |
| | *Index note* | *309* |
| | *Table of clause references* | *310* |

# Preface

I had no plans to write this book and it is with some reservations that I now do so. The New Engineering Contract aims for a new order in construction and its objectives can only be applauded. It will present few problems to users determined that it should succeed and it has already acquired a loyal band of followers who herald it an improvement on other standard forms of contract.

Is it necessary, therefore, to subject the New Engineering Contract to scrutiny when faith in its principles may be more important than the detail of its provisions? At first I thought not. Such is the novelty of style of the New Engineering Contract and so dependent is it on the concept of co-operation, that I doubted it could ever attract widespread interest in the conservative and combative construction industry.

But Sir Michael Latham changed all that. With his glowing endorsement of the New Engineering Contract in his report *Constructing the Team*, Sir Michael ensured that if the contract fails to find a place in the top rank of construction contracts it will not be for lack of attention. The New Engineering Contract may not yet be the most widely used contract in the land but it is certainly the most talked about. And despite the reservations of some eminent lawyers and the cautions voiced by some very experienced engineers, use of the contract is accelerating rapidly.

There can be only one outcome to this. Once the New Engineering Contract is in general use it will be put under the microscope like any other construction contract. Its good intentions will not preserve it from confrontation with commercial self interest. The question which will determine whether or not the New Engineering Contract will succeed in general use is not whether or not its aims are laudable but whether it is robust enough to withstand detailed analysis and criticism.

In writing this book I make absolutely no claim to having a better understanding of the New Engineering Contract than anyone else. I have simply set out my thoughts and queries with the intention of stimulating debate and alerting users of the contract to its potential difficulties.

I dedicate this book to all those who have so kindly shared with me their experiences in early use of the contract.

April 1996

Brian Eggleston
5 Park View, Arrow
Alcester
Warks B49 5PN

# Author's note

### Phraseology

The New Engineering Contract is a family of contract documents and the proper use of the acronym NEC is as a prefix rather than as the name of any single contract. This book is principally a commentary on the NEC Engineering and Construction Contract – the main contract in the family – which in time may become known as the ECC. However, at present the Engineering and Construction Contract is still generally referred to as the NEC and that is the policy adopted throughout this book.

### Capitals

The NEC relies heavily on defined terms which have capital initials and identified terms which are in italics. However, for reasons of style which I hope make for easier reading, capitals and italics have been used sparingly in this book, and therefore both defined terms and identified terms appear usually in ordinary lower case.

### Text of the NEC

Very little of the text of the NEC is quoted in this book. I have assumed that readers will have to hand a copy of the NEC and the other forms in the family as appropriate.

Generally, commentary on the text relates to the second edition of the NEC (date stamped November 1995) but for the professional services contract and the adjudicator's contract first editions are still in force.

# Chapter 1

# Introduction

## 1.1 Background

The New Engineering Contract system is an interlocking family of contracts published for the Institution of Civil Engineers by Thomas Telford Services Ltd. It is wholly original in its style and courageously ambitious in its intentions.

**The NEC family of contracts**

The contracts currently (as at January 1996) in the NEC family comprise:

- the Engineering and Construction Contract
- the Engineering and Construction Subcontract
- the Professional Services Contract
- the Adjudicator's Contract.

Other contracts which are expected to be added in due course include:

- the Products Contract
- the Maintenance Contract
- the Minor Works Contract.

**Publication**

The main contract and the subcontract were first published as a consultative edition set in January 1991. The formal first edition followed in March 1993 and the second edition in November 1995.

First editions of the Professional Services Contract and the Adjudicator's Contract were published in 1994. As at April 1996 neither has been superseded.

**Supplementary documents**

Officially published within the NEC system are guidance notes and flow charts. Neither are to be taken as contract documents and both explicitly state that they should not be used for legal interpretation of the contract documents.

## 1.2 Aims and objectives

Development of the NEC first started in 1986 as a fundamental review of alternative contract strategies with the objective of identifying the needs for good practice. Behind the review were the beliefs that existing forms of contract did not adequately serve the best interests of the parties and that contracts can be designed to promote good management and to reduce the incidence of disputes.

The three specific objectives which guided the drafting team of the new contract were:

- that it should be more flexible in its scope than existing standard forms
- that it should provide a greater stimulus to the good management of projects than existing forms
- that it should be expressed more simply and clearly than existing forms.

**Flexibility**

Flexibility is perhaps the most ambitious of these objectives. The NEC aims to be an all-purpose contract for all construction and engineering disciplines at home or abroad. It offers this through a combination of uniquely drafted provisions and a complex structure of options. Four distinct features are presented:

- the NEC avoids discipline specific terminology and references to the practices of particular industries. It relies instead on a framework of general provisions written largely in non-technical language
- responsibility for design is not fixed with either the employer or the contractor but can be set at any amount from nil to total with either party
- six primary options give a choice of pricing mechanisms from lump sum to cost plus
- 15 secondary options allow the employer to build up the provisions in the contract to suit his individual policies.

**Stimulus to good management**

As mentioned above much of the inspiration for the development of the

NEC stems from a belief that existing forms of contract no longer adequately serve the best interests of the parties. The argument is put that expanding procurement strategies, changing practices in contracting, and developments in project management require contracts to focus as much on management as on the obligations and liabilities of the parties. So the NEC lays great emphasis on communications, co-operation and programming and the need for clear definition at the outset of various types of information.

Reports from early users of the NEC suggest that improvements in project management are being achieved and that job satisfaction for those involved is better than with traditional contracts. But most early users of the NEC admit that they are either enthusiasts or willing pioneers and it is probably too early to say how the NEC will fare in general practice.

**Clarity and simplicity**

The approach adopted by the drafting team towards the objective that the NEC should be expressed more simply and clearly than existing forms of contract was to start from scratch rather than to build on old foundations. So the NEC is intentionally and conspicuously different from other standard forms in its style and its structure.

It is written in non-legalistic language using short sentences and avoiding cross-references. Familiar phrases such as 'extension of time' and 'variations' are absent as is the regular use of the word 'shall' to signify obligations. However, there is a price to pay for this brevity. Taken by itself the NEC is, at least for first time readers, so novel as to be more of a mystery than a model of clarity and simplicity. Fortunately the guidance notes and flow charts are there to assist in general understanding and the application of the contract.

But for legal interpretation of the contract the problem is not so easily solved. Neither the guidance notes nor the flow charts are intended to be used for legal interpretation and the application of legal precedents from traditional forms of contract written in conventional drafting style can only be surmised; which raises the question, has the NEC sacrificed legal certainty in pursuit of a new order?

There are certainly some who feel that discarding conventional drafting amounts to discarding the accumulated contractual wisdom of generations. Throwing the baby out with the bathwater is how one eminent construction lawyer has put it. But others are far more optimistic and they suggest that to focus on the words of the NEC is to miss the point of its message; and that the courts, if called upon to do so, will have no difficulty in discovering the true intentions of the parties.

The difficulty with the latter approach is that it tends to overlook a basic rule in the construction of contracts: that the courts in seeking to give effect to the intention of the parties will give effect to the intention as expressed by ascertaining the meaning of the words actually used.

## 1.3 Structure of the NEC

Each NEC contract is uniquely put together to meet the employer's needs by assembling clauses from the option structure and by particularisation in accompanying documents.

**The option structure**

The employer:

- makes a selection from the six main options as to which type of pricing mechanism is to apply
- includes in the contract the nine sections of core clauses
- includes in the contract such selection (if any) from the 14 detailed secondary option clauses as he thinks fit
- includes in the contract under the fifteenth secondary option any additional clauses required by him or as agreed with the contractor.

*The main options*
The main options comprise six types of payment mechanism:

- Option A – priced contract with activity schedule
- Option B – priced contract with bill of quantities
- Option C – target contract with activity schedule
- Option D – target contract with bill of quantities
- Option E – cost reimbursable contract
- Option F – management contract.

Each of the main options is published in a separate book which includes the relevant core clauses for the particular option.
Note that there is no main option for construction management and none specifically for design and build.

*The core clauses*
The core clauses are grouped into nine sections, numbered as follows:

(1) General
(2) The contractor's main responsibilities

## 1.3 Structure of the NEC

(3) Time
(4) Testing and defects
(5) Payment
(6) Compensation events
(7) Title
(8) Risks and insurance
(9) Disputes and termination.

For each section there is a common set of core clauses and for some of the main options there are additional core clauses.

*The secondary options*
The 15 secondary option clauses are labelled G to Z. Included within them are some matters such as retention and liquidated damages for late completion which most traditional contracts treat as essential. The full list is:

- Option G – performance bond
- Option H – parent company guarantee
- Option J – advance payment
- Option K – multiple currencies (for use with Options A and B)
- Option L – sectional completion
- Option M – limitation of design liability
- Option N – fluctuations (for use only with Options A, B, C and D)
- Option P – retention (for use only with Options A, B, C, D and E)
- Option Q – bonus for early completion
- Option R – delay damages (liquidated)
- **Option S – low performance damages**
- Option T – changes in the law
- Option U – special conditions
- Option V – trust fund
- Option Z – additional conditions.

For Option Z (additional conditions) the promoters of the NEC recommend that these are written as far as possible in the style of the NEC. There is an interesting legal point on additional conditions discussed further in Chapter 3, section 1, on whether any special conditions of contract included in the contract through Option Z follow the usual rule that special conditions of contract take precedence over standard conditions in the event of ambiguity.

**Accompanying documents**

The NEC does not define the 'contract'. Clearly the schedules of cost components which are printed in with the contract are incorporated by reference if not by the fact of their location. Similarly the contract data sheets which allow the employer and the contractor to state particulars relating to the contract must have contractual effect.

Two further key documents, or sets of documents, which are fundamental to the NEC but which are wholly particular to each contract are the works information and the site information. Providing these are properly identified in the contract data they become contract documents by reference. The position is similar in respect of activity schedules and bills of quantities.

It would probably be going too far to say that the NEC makes the accepted programme a contract document. But the accepted programme is certainly an accompanying document in the sense that it is of significant contractual effect with regard to the employer's obligations and the contractor's financial entitlements.

## 1.4 Special characteristics of the NEC

As will be seen from later chapters in this book the NEC has many special characteristics, some obviously included by design, others present perhaps by accident.

### The Latham principles

By design the second edition of the NEC aims to comply with all 13 principles listed in Sir Michael Latham's report *Constructing the Team* as essential characteristics for a modern form of contract:

(1) A specific duty for all parties to deal fairly with each other, and with their subcontractors, specialists and suppliers, in an atmosphere of mutual co-operation.

(2) Firm duties of teamwork, with shared financial motivation to pursue those objectives. These should involve a general presumption to achieve 'win-win' solutions to problems which may arise during the course of the project.

(3) A wholly interrelated package of documents which clearly defines the roles and duties of all involved, and which is suitable for all types of project and for any procurement route.

(4) Easily comprehensible language and with guidance notes attached.

(5) Separation of the roles of contract administrator, project or lead manager and adjudicator. The project or lead manager should be clearly defined as client's representative.

(6) A choice of allocation of risks, to be decided as appropriate to each project but then allocated to the party best able to manage, estimate and carry the risk.

(7) Taking all reasonable steps to avoid changes to pre-planned works information. But, where variations do occur, they should be priced in advance, with provision for independent adjudication if agreement cannot be reached.

(8) Express provision for assessing interim payments by methods other than monthly valuation, i.e. milestones, activity schedules or payment schedules. Such arrangements must also be reflected in the related subcontract documentation. The eventual aim should be to phase out the traditional system of monthly measurement or remeasurement but meanwhile provision should still be made for it.

(9) Clearly setting out the period within which interim payments must be made to all participants in the process, failing which they will have an automatic right to compensation, involving payment of interest at a sufficiently heavy rate to deter slow payment.

(10) Providing for secure trust fund routes of payment.

(11) While taking all possible steps to avoid conflict on site, providing for speedy dispute resolution if any conflict arises, by a predetermined impartial adjudicator/referee/expert.

(12) Providing for incentives for exceptional performance.

(13) Making provision where appropriate for advance mobilisation payments (if necessary, bonded) to contractors and subcontractors, including payments in respect of off-site prefabricated materials provided by part of the construction team.

**The NEC provisions**

Without considering here the detail of whether the provisions of the NEC match its principles – that follows in later chapters – it is worth noting two general points which distinguish the NEC from other standard forms.

The first is that the NEC cannot be taken literally. A good many of its provisions only work if interpreted and applied with common sense. Other provisions, if applied strictly to the letter, would be grossly unfair – usually to the contractor.

The second point is that the NEC is not a contract to be filed away at commencement and dusted down when problems arise. The NEC is as much a manual of project management as a set of conditions of contract. It demands constant attention, a high level of administrative support and, judging from reports on current NEC contracts, a large bank of filing cabinets.

Chapter 2

# The main options

## 2.1 Introduction

Flexibility, as discussed in Chapter 1, is a key objective of the NEC. In practice it is already proving to be one of the contract's key attractions. No other standard form of contract permits such a wide range of contract strategies to be applied, and none competes with the NEC in offering common core clauses for contracts as diverse in their pricing arrangements as lump sum to cost plus.

The highly visible aspects of the NEC's flexibility are that decisions have to be made at the outset on which main option is to be used and which secondary options are to apply. Less visible, but certainly important to the choice of the main option, is which party is to be responsible for design, either in whole or in part. So although it might appear at first sight that the choice between the six main options is simply one of pricing mechanism, the reality is that the choice, if it is to be made correctly, requires a sophisticated analysis of the particulars of the project and the requirements of the employer.

**The main options**

The six main options of the NEC are:

- Option A – priced contract with activity schedule
- Option B – priced contract with bill of quantities
- Option C – target contract with activity schedule
- Option D – target contract with bill of quantities
- Option E – cost reimbursable contract
- Option F – management contract.

These options provide, in descending order, a broad scale of distribution of risk in price with Option A providing maximum certainty of price for the employer and Option F providing the least.

The employer is required to state in part one of the contract data which main option is to be used and in most cases the choice of main option will be entirely with the employer. But sometimes potential tenderers will be invited to propose which main option should apply as part of pre-qualification procedures. And obviously where partnering is intended the prospective contractor can be expected to have a say in which main option should be used.

First time users of the NEC should be alert to the fact that each main option has its own particular clauses which are additional to the core clauses in the nine sections of the NEC. In particular, a point to note is that although the definitions in the core clauses stop at number 17 in Section 11 of the base contract, they continue to number 30 in some of the main options.

**Construction management**

There is no named main option in the NEC for construction management – the system in which the contractor provides only management services to the employer with the works packages let as contracts directly between the works contractors and the employer. However, this need not be a barrier to the use of the NEC for construction management.

The promoters of the NEC suggest, and there is no obvious reason why it should not be possible, that for construction management the employer should appoint a construction management contractor as project manager under the NEC professional services contract. The duties of the construction manager would then be to advise the employer on the placing of the works contracts under whichever main options of the NEC were most appropriate, and then the project management of the works contracts.

## 2.2 Contract strategy

Contract strategy is not an exact science. Some guiding principles certainly apply but every employer is unique in his aspirations, his circumstances and his preferences.

For some employers certainty of price is the dominant aspiration and then, given few restrictive circumstances and particular preferences, the obvious strategic choice will be a lump sum contract with contractor's design. But for other employers certainty of price may be secondary to considerations of quality, operations/restrictions, or the need for a quick start and a fast finish. Which method of procurement? Which type of contract? Which form of contract? These then become more complex questions. And, of course, some employers, on the strength of past experiences or hopes for the future, develop preferences for certain methods of procurement and certain forms of contract. Rational analysis of selection criteria to determine contract strategy may then become secondary to selection of the most suitable contractor.

One of the strengths of the NEC is that if the employer does develop a preference for its use he is nothing like as limited in his choice of procurement route as with other standard forms. He has six main options to choose from and construction management available as a further option.

It is not appropriate in this book to provide a detailed review of the theories of contract strategy but for those who do need to study the subject a useful starting point is CIRIA Report R85 on Target and Cost-reimbursable Construction Contracts or the RIBA publication *Which Contract*.

As a checklist for matters to consider the following may be helpful:

- which party is to be responsible for design
- how important to the employer is certainty of price
- what views prevail on the allocation of risk
- how firmly known are the employer's requirements and what likelihood is there of change
- what operating restrictions apply on the employer's premises or in the construction of the works
- what emphasis is to be placed on early commencement and/or rapid completion
- what flexibility does the employer need in the contractual arrangements, e.g. to terminate at will
- how anxious is the employer to avoid or to minimise formal disputes and legal proceedings
- how important to the employer is the concept of single point responsibility.

**Responsibility for design**

The general principle which should influence which party is responsible for design is that of competence. Which party can most competently undertake the design?

If professional design firms are to be employed, whether it be by the employer or by the contractor, the question of competence in this general sense does not arise. But with contractor's design an obvious advantage for the employer is that a choice of designs may be put forward by the tenderers. A further potential advantage is that the contractor's expertise is more likely to be used to the full when the freedom to develop that expertise in the design is permitted.

If the employer already has his own in-house design resources it may be neither efficient nor economic to place design responsibility with the contractor. And it may well be that in-house design teams are more closely in tune with the employer's requirements than any contractor could be.

In some situations there are matters of confidentiality as to the purpose or operation of the works, which are wholly decisive as to whether design briefs can be issued to tenderers and as to which party is responsible for design. In other situations there may be a reliance on specialist know-how or patented designs, which is itself decisive as to design responsibility.

But as a general rule, if the employer is able to specify his requirements in terms of a performance specification or quality standards there is much to be said for contractor's design. Not only may the standard of liability of a contractor for his design (fitness for purpose) be higher than that of a professional designer (skill and care) but the scope for claims from the contractor for extra payment arising out of the designer's defaults and deficiencies is eliminated.

As to how the allocation of responsibility for design influences choice between the main options of the NEC, the main points to note are:

- Option A – lump sum contract
  Ideally suited to contractor's design but can be used for employer's design or divided design responsibility providing the employer's design element is complete at the time of tender.

- Option B – remeasurement contract
  Not suited to contractor's design because of the reliance on bills of quantities and the difficulties posed by the contractor producing his own bills of quantities.

- Option C – target contract (lump sum base)
  As Option A but allows the employer more flexibility in developing his own design.

- Option D – target contract (bill of quantities base)
  Suffers from similar problems to Option B.

- Option E – cost reimbursable contract
  Permits maximum flexibility in allocation of design responsibility and allows development of the design as the works proceed.

- Option F – management contract
  Not suitable for allocation of the whole of design responsibility to the contractor unless placed as a 'design and manage' contract. But is particularly suitable for contracts with a high reliance on specialist subcontractors who undertake their own design.

## Certainty of price

For many employers certainty of price is the decisive factor in contract strategy. Commercial pressures may dictate that either a project can be completed within a set budget or it is not worth commencing.

Option A (the lump sum contract) offers the best prospects for certainty of price, particularly when used with contractor's design.

Option C (the target contract based on lump sum) fixes with some degree of certainty the maximum price but at tender it is less precise than Option A in fixing the likely contract price.

Options B and D (both bill of quantities based) put the risk of accuracy of quantities on the employer and consequently both suffer from lack of price certainty.

Option E (the cost reimbursable contract) relieves the contractor of any risk on price (other than in his fee). Consequently not only is the employer at risk on the price, with the contract itself providing no certainty of price, but the contractor has little incentive by way of any target to minimise costs. Clearly Option E is not suitable if the employer is looking for certainty of price.

Option F (the management contract) is a cost reimbursable contract in so far as the employer and not the contractor takes the risk on the costs of the works contracts. However, management contracts are frequently arranged on the basis of lump sum works contracts and this can introduce a good measure of cost control into the system. If the quotations for the works contracts can all be obtained before the letting of the management contract there can also be a good measure of price certainty.

**Allocation of risk**

The guiding principle on allocation of risk is that risk should be allocated to the party best able to control it. Most contracts, including the NEC, show some regard for this principle but few, and the NEC is no exception, take it to its ultimate conclusion. Two other factors frequently prevail.

One is that it is better for the employer to pay for what does happen than for what might happen – hence, unforeseen ground condition clauses. The other is that in the interests of fairness (and in some cases coincidental commercial interests) the contractor should not be required to carry risks which are uninsurable or which arise from matters beyond the influence of either party – for example, changes in statute which affect the costs of construction.

Taken together, the result of the above is that the employer can often end up carrying some risks over which he has no control whatsoever. Thus if the government puts up labour taxes the employer usually pays the additional contract costs although it is only the contractor who has any control over those costs.

When it comes to the selection of a main option of the NEC the employer is fully justified in asking how the various options deal with the allocation of risk. The answer, surprisingly perhaps, is that apart from the variables inherent in the pricing mechanisms of the main options and the variables which can be introduced through choice of secondary options, the NEC operates a policy of common allocation of risk through all its main options. It does this quite deliberately to provide consistency in the application of its core clauses and its compensation events. But it is certainly questionable whether the employer's interests are always best served by the policy. For example, is it appropriate that target cost prices should be adjustable for the full range of compensation events? And is there a proper place for unforeseen ground conditions clauses in design and build contracts?

The answers to these questions are not wholly academic even if employers desist, as they are encouraged to do by the promoters of the NEC, from making changes to the core clauses and the set list of compensation events to suit their particular projects. What employers need to do is to take note of the common aspects of allocation of risk in the main options and to consider what influence that should have on contract strategy generally.

So, as an example, an employer wishing to develop a difficult site with uncertain ground conditions might well decide – returning to the principle that risk should be allocated to the party best able to control it – that retaining responsibility for design would be more appropriate than contractor's design and that Option B might be more favourable than Option A in obtaining competitive tenders.

### The employer's requirements

The aspects of the employer's requirements which influence the selection of the main option of the NEC are various. They include:

- the degree of finalisation of the requirements
- the likelihood of change in the requirements
- the extent to which the requirements are performance based
- the extent to which the requirements are confidential
- the extent to which the requirements involve active participation of the employer in the construction of the works.

As far as finalisation of the requirements and the likelihood of change are concerned, the simple rule of contract procurement is that you should only buy on a lump sum basis when you know in advance what you want. Changes and variations are likely to be expensive and associated claims for delay and disruption even more so. Not uncommonly the apparent certainty of the lump sum price evaporates as changes, variations and claims are paid on cost plus. It makes good commercial sense, therefore, for employers who know they are likely to end up paying cost plus to embark on a cost plus contract in the first place. They will then have some control over the costs from the outset and they can consider whether a target price contract is appropriate, so as to provide the incentive for all costs to be minimised.

Options A and B of the NEC, being firm price contracts, are clearly least suited to change and/or development of the employer's requirements as the works progress. Options E and F, being cost plus contracts, are clearly best suited. They allow the employer maximum flexibility.

The two target contract main options, C and D, provide an intermediate level of choice. They do allow flexibility but they require a reasonable level of definition of the employer's requirements at the outset in order for target prices to be set.

Performance criteria, confidentiality matters and employer participation have much to do with decisions to be made on allocation of design responsibility as discussed above. But taken separately, so far as that is possible:

- the ideal choice for a performance contract would be Option A
- the necessary choice for maximum confidentiality may be Option F
- the appropriate choice for employer participation is probably Option E.

**Operating restrictions**

In contracts where there are significant operating restrictions on the contractor either because of the location of the site or because parts of the site contain continuous production facilities or the like, the essential question for contract choice is how well the restrictions can be defined in the tender documents. A secondary question is whether or not the restrictions are likely to be subject to change.

If complete definition of restrictions is possible at tender stage there is no reason why Options A and B should not be used, however onerous the restrictions. But if complete definition is not possible, or change is likely, then Options A and B are not suitable because of their inherent inflexibility.

**Early start and/or rapid finish**

Timing requirements have much to do with the selection of the best main option for any particular contract.

Options A and B, requiring the maximum definition of detail at tender stage, have the longest lead times. Options E and F which can commence with minimum definition have the shortest lead times. Options C and D occupy the intermediate position.

As to completion times and how rapidly a finish can be achieved, that comes down mainly, in consideration of the main options, as to how well each permits development of the design as the works proceed. That apart there is not much to choose between the options, except possibly that with the cost reimbursable options the employer has greater flexibility in ordering acceleration.

**Flexibility in contractual arrangements**

As a general rule the employer has more flexibility under cost reimbursable contracts to change not only the detail of his technical requirements but also the detail of contractual arrangements. This follows naturally from the payment mechanism.

One of the more evident, and perhaps one of the most important aspects of this flexibility, is whether there is the facility for the employer to terminate the contract at will, without any suggestion of fault on the part of the contractor. The NEC has an elaborate scheme in the Section 9 core clauses for dealing with termination and the amounts due on termination. It permits termination at will for all the main options.

**Avoidance of disputes**

It may seem odd that with a contract such as the NEC, committed to the cause of avoidance of disputes, it can be suggested that the employer's desire to avoid disputes should find its way into the selection procedure for one of the main options. Surely, it might be said, all are equally non-adversarial.

But, in reality, that is not the case. All the main options have common core clauses and a common set of compensation events, but that does not stop firm price options A and B being potentially more adversarial than the cost reimbursable options E and F. Nor does it alter the fact that design and build contracts give the contractor fewer opportunities for making claims than employer designed contracts.

Consequently, if the avoidance of disputes is particularly important to the employer, that should be a factor taken into account in the early stages of contract strategy. And it is wholly appropriate that the employer should select the main option with a view to minimising the use by the contractor of the compensation events.

**Single point responsibility**

For some employers the concept of single point responsibility is important enough to influence their entire contract strategy. Principally the matter is one of allocation of design responsibility which in turn works its way into selection of the appropriate main option of the NEC. So, to take the simplest example, the employer who contracts on a turnkey basis (turn the key and everything is done) will select a design and build contractor, will specify what he requires in performance terms, and will choose Option A.

## 2.3 Option A – priced contract with activity schedule

Option A is described as a priced contract with activity schedule. There is no definition of Option A beyond this in the contract so to understand what Option A is, and how it differs from the other five main options of the NEC, it is necessary to look at the clauses of the NEC applying particularly to Option A. In total there are ten such clauses but for the purpose of defining Option A three are particularly important:

- clause 11.2(20) – the prices are the lump sum prices for each of the activities in the activity schedule unless later changed in accordance with the contract
- clause 11.2(24) – the price for the work done to date is the total of the prices for each group of completed activities and each completed activity which is not in a group
- clause 54.1 – information in the activity schedule is not works information or site information

Turning now to Section 5 (payment) of the NEC, clause 50.2 states that the amount due to the contractor is the price of the work done to date.

So what can be gathered from the above is that Option A is a lump sum contract in which the lump sum price is broken down into subsidiary lump sum prices for the various activities to be undertaken in providing the works. There is nothing unusual in this in that a lump sum contract price, whatever the form of contract, is usually supported by a breakdown of the contract price in the form of a schedule to be used either for making interim payments or assisting in the valuation of variations. The difference in the NEC is that there is no definition of the contract price and no specific statement to the effect that the contractor's obligation is to provide the works for the contract price.

And to further emphasise the significance in the NEC of the lump sum prices for activities, changes of prices resulting from the assessment of compensation events are made as changes to the prices of activities.

The legal effects of this are difficult to assess. Perhaps much will depend in any particular case on how the form of tender is worded – there is no standard form. But there must be the possibility that if the form of tender follows too closely the wording of the NEC and puts the contractor's offer in terms of lump sum prices for activities, then the contract may be held to be for a series of lump sum prices rather than for a single lump sum. Perhaps this is what the NEC intends, although it is far from obvious what advantage accrues.

But since most parties using Option A of the NEC will normally intend to contract on the basis of a single lump sum price for the works, it is probably best, for the avoidance of doubt, that the form of tender avoids any confusion and states clearly that the contractor's offer is for a single lump sum.

**The activity schedule**

The activity schedule is not a defined term of the NEC. It is mentioned many times throughout the contract but the only clear indication of what it is in contractual terms is given in part two of the contract data. There it says that if Option A or C is used the activity schedule 'is...' and a space is provided for the contractor to state what the activity schedule 'is'.

In practice most users of the NEC will understand that the activity schedule is a breakdown of the work to be done under the contract. What may not be so obvious is that under the NEC the activity schedule must cover the whole of the contract price, and the contractor's entitlement to interim payments is assessed on the basis of completed activities.

Contractors using the NEC have soon learned the lesson that the more activities they list the more regular are their interim payments. Thus, to list a bridge abutment as an activity allows interim payment only when the abutment is wholly completed. But broken down into excavation, piling, blinding concrete, formwork, reinforcement, concrete placing, concrete finishing etc., interim payments become due for each completed operation.

One result of this is that activity schedules running into hundreds, if not thousands, of items are being produced with consequent effects on programmes which, by clause 31.4, must show the start and finish of each activity on the activity schedule. And since the assessment of compensation events requires changes to the activity schedule and correspondingly the programme, the administrative burden is obvious. In some cases the amount of detail is so great that only computers can keep pace with the changes to the activity schedule and the programme. To counter this some employers have taken to fixing themselves the particular activities which can be listed in the activity schedule, either by directions in the instructions to tenderers or by amendment of the contract data.

## 2.4 Option B – priced contract with bill of quantities

Option B of the NEC which is described as a priced contract with bill of quantities is what is traditionally known as a remeasurement contract. So much can be determined from:

- clause 11.2(21) which states that the prices are the lump sums and the amounts obtained by multiplying the rates by the quantities for the items in the bill of quantities, *and*
- clause 11.2(25) which states that the price of the work done to date is the total of the quantity of work completed for each item in the bill of quantities multiplied by the rate plus a proportion of each lump sum as so completed.

The combination of terminology 'lump sums' and 'rates' is unusual and care may have to be taken in drafting the form of tender to avoid the possibility of confusion as to the construction of the contractor's offer.

**Method of measurement**

The NEC does not refer to any particular method of measurement. It relies on the employer stating in part one of the contract data which method of measurement is used.

## 2.5 *Target contracts generally*

Target price contracts are versions of cost reimbursable contracts where the reimbursement of cost ceases or reduces when a target price is reached.

Contracts where reimbursement ceases altogether at the target price are sometimes called GMP contracts (guaranteed maximum price).

For most target contracts, however, a sliding scale of reimbursement operates both above and below the target price so that the employer and the contractor share the financial risks. The contractor, in effect, gains a bonus if he can keep the actual cost below the target price, but he shares the cost when the actual cost exceeds the target price. So target price contracts encourage the contractor to be efficient in the use of resources.

For contractors, however, there is a real danger that sight can be lost of the financial risks of target contracts. Because reimbursement is on a cost plus basis at the outset and remains that way for much of the contract, too little attention may be given to the impending effects of cost over-runs.

**Target setting**

Target price contracts can be used with either contractor's design or employer's design but whichever applies there must be a reasonable definition of the employer's requirements at tender stage to enable the tenderers to reach their assessments of the target price. In some cases a

## 2.5 Target contracts generally

performance specification alone is sufficient but in other cases drawings and indicative bills of quantities are supplied by the employer.

It is not unusual for protracted negotiations to take place, before the award of a target price contract, on the precise figure at which the target should be set. Obviously it is in the contractor's interests to secure the contract at the highest achievable target price.

### Competition

Competition operates between tenders in target price contracts in two ways:

- between the fees tendered to cover non-reimbursable costs – principally overheads and profit. The fees are usually tendered on a percentage basis (to be added to reimbursable cost) but they may be lump sums
- between the target prices tendered reflecting the assessments of the various tenderers on final actual cost.

In comparing tenders employers use various formulae to analyse the balance between the different levels of tender fees and target prices, but it is not unknown for employers to fix either the fees or the target prices to simplify comparisons.

### Target price adjustment

For contractors embarking on target price contracts a key question is how restrictive (or how generous) are the permitted adjustments to the target price once the contract is in operation. Clearly, at the very least, there must be upward adjustment for changes and variations which require additional works, otherwise the employer might receive the benefit at no cost. But what of such matters as unforeseen ground conditions or other unexpected costs? That depends on the policy of risk allocation in the contract – and the NEC is fairly generous in that all the compensation events can adjust the target price.

One advantage of the NEC target price contracts (Options C and D) over some other target contracts in use is that they are clear on their policies for target price adjustment. Contractors should beware of straightforward cost reimbursable contracts applied to target price contracts. It is necessary to see what amendments have been made to cover target price adjustments. The standard IChemE Green Book, for example, says nothing on unforeseen ground conditions, and does not need to, since all costs are reimbursable. Without some amendment for this in a target price contract the result can be that the contractor ends up taking risks which he never contemplated and were never apparent.

### Risk sharing formulae

The simplest arrangement for risk sharing above and below the target price is that each party bears 50% of any cost over-run and each takes 50% of any saving. Most target price contracts, however, have more sophisticated arrangements with sliding scales of risk distribution. Not infrequently there is a cut off point for cost reimbursement at 15% or so above the target price, which effectively creates a guaranteed maximum price (subject only to target price adjustments).

The NEC provides for the employer to enter in part one of the contract data various share percentages against a range of percentage changes from the target price, whenever either Option C or Option D is used.

### Disallowed costs

Even in straight cost reimbursable contracts there are usually some items of cost which are disallowed, either because they arise from some specified default or breach on the part of the contractor or because they are not properly substantiated. The contract will normally list those items which are to be regarded as disallowed costs.

With target cost contracts the lists of such items are sometimes more extensive than those for straight cost reimbursable contracts. But this is an area where policies of contracts (and employers) differ considerably, particularly on the question of whether the costs of rectifying defects should be reimbursable or should be disallowed.

The NEC applies a common list of disallowed costs to its target cost contracts (Options C and D) and to its cost reimbursable contract (Option E). It does, however, have a more restricted list for its management contract (Option F).

### Payment arrangements

The control of costs in cost reimbursable contracts can be extremely complex and time consuming. The amount of paperwork to be processed can be enormous. This is recognised in the IChemE Green Book where interim payments each month are made on a combination of estimated costs and incurred costs.

The payment arrangements in the NEC, however, are much the same for cost reimbursable contracts as for firm price contracts. The project manager is required to assess the amount due and to certify the same within one week of each assessment date (clause 51.1). The amount due is the price for work done to date.

For Options C and D the price for work done to date is the actual cost the contractor has paid plus the fee. By any standards the assessment of this within one week is an ambitious task.

## 2.6 Options C and D – target contracts

Option C is described as a target contract with activity schedule; Option D as a target contract with bill of quantities. The only significant differences between the two are:

- in Option C the target price is based on a lump sum (split into activities) whereas in Option D the target price is based on a bill of quantities
- in Option D the employer takes the risks on changes of quantities (and departures from the method of measurement) and the target price is adjusted according to the final measure.

Surprisingly, nowhere in the NEC is the phrase 'target price' actually used. 'The Prices' as defined in clause 11.2 (lump sums for activities; rates for quantities for bills) are apparently to be taken in Options C and D as the target price. This may just about be a workable arrangement but it is not particularly satisfactory since the target mechanism is obviously never intended to apply individually to either activities or rates and quantities.

The purpose of the activity schedule in Option C and the bill of quantities in Option D is different from the purpose of those documents in Option A and Option B. For Options C and D the documents do not fix amounts due as interim payments. They serve only in the assessment of compensation events (which move the target price) and in calculation of the contractor's share.

**Points to note**

It is not intended that this chapter should cover in full detail the operation of the core clauses and the particular clauses applying to Options C and D, but the following common points are worth noting:

- Clause 11.2(23) – which states that the price for the work done to date is the actual cost which the contractor has paid plus the fee. The importance of this is that under clause 50.2 the amount due on interim payments is the price for the work done to date. The intention seems to be that under Options C and D (the same applies to Option E but not to Option F) the contractor is only paid for that which he has already paid out himself. The financing implications of this on the contractor are only too obvious.

## 2.6 Options C and D – target contracts

- Clause 11.2(27) – which defines the meaning of the term 'Actual Cost' in Options C, D and E. The definition allows for the recovery of payments to subcontractors and is quite different from the definition of 'Actual Cost' applying to Options A and B.
- Clause 11.2(30) – the costs of correcting defects for which the contractor is responsible are disallowed.
- Clause 20.3 – the contractor has a duty to advise the project manager on practical implications of the design and on subcontracting arrangements.
- Clause 20.4 – the contractor is required to prepare regular forecasts of total actual cost.
- Clause 36.5 – the contractor is obliged to submit any subcontractor's proposal to accelerate to the project manager for acceptance. The intention of this is not obvious but its effect appears to be to take full control over the pace of the works away from the contractor.
- Clauses 53.3 and 53.4 – the contractor's share of any saving or excess relative to the target price is assessed firstly on completion of the whole of the works and again when the final account is known. It appears that the contractor continues to be paid up to completion on a cost plus basis even when the costs exceed the target price. The contractor's share of the excess is apparently refunded to the employer only after completion. In the event of a serious cost over-run this could embarrass an employer with limited financial resources.
- Clause 53.5 – proposals by the contractor to change the works information provided by the employer so as to reduce actual cost do not lead to reduction in the target price. Thus, the contractor takes his share of any savings from timing changes/variations which he initiates.

## 2.7 Option E – cost reimbursable contract

Cost reimbursable contracts, as will be apparent from comments earlier in this chapter, put the least financial risk on the contractor and give the employer the least certainty of price. Their grave defect is that they provide no incentive for the contractor to minimise costs and, when the contractor's fee is on a percentage basis, they encourage expenditure. Not surprisingly, therefore, cost reimbursable contracts tend to be used only as a policy of last resort and in circumstances when other procurement methods are not appropriate.

Option E of the NEC is a straightforward cost reimbursable contract which operates on actual cost plus the percentage fee inserted in part two of the contract data by the contractor.

## 2.7 Option E – cost reimbursable contract

The comparison of tenders on financial grounds is principally on the various tendered fee levels. Note should, however, also be taken of the particular rates for those parts of the schedule of cost components which tenderers are required to price in part two of the contract data.

**Points to note**

Particular points to note in Option E are much the same as those listed above for Options C and D, except that the target related points have no application in Option E.

## 2.8 Option F – management contract

A management contract is a form of cost reimbursable contract where the contractor subcontracts all or most of the construction works. In Option F of the NEC clause 20.2 requires that the contractor shall subcontract:

- design
- construction
- installation
- other work

as stated in the works information. It permits the contractor to subcontract or carry out work himself which is not so stated.

**Payment**

Option F differs from Options C, D and E in that payment is related to the actual cost which the contractor has accepted for payment and is not fixed by the amount paid – clause 11.2(22).

**Other points to note**

- Clause 11.2(26) – the definition of actual cost covers only subcontracted work and it appears to exclude the cost of any work undertaken directly by the contractor.
- Clause 11.2(29) – disallowed cost is different under Option F from under Options C, D and E. The costs of rectifying defects are not expressly disallowed.
- Clause 20.2 – the contractor's obligations in providing the works are to manage:

○ the contractor's design
○ the construction and installation of the works.

This suggests that there is no obligation on the contractor to undertake any design or construction work himself (as distinct from managing it).

# Chapter 3

# The secondary options

## 3.1 Introduction

The core clauses of the NEC deal only with obligations and procedures considered essential to all contracts. By themselves they will rarely be sufficient to cover all the contractual detail for any particular project. For example, the core clauses have nothing to say on performance bonds, retentions or liquidated damages.

To cover additional detail the NEC has 14 secondary option clauses, lettered G to V, all drafted to be compatible with the core clauses. They are:

- G – performance bond
- H – parent company guarantee
- J – advanced payment to the contractor
- K – multiple currencies
- L – sectional completion
- M – limitation of the contractor's liability for design to reasonable skill and care
- N – price adjustment for inflation
- P – retention
- Q – bonus for early completion
- R – delay damages
- S – low performances damages
- T – changes in the law
- U – the Construction (Design and Management) Regulations 1994
- V – trust fund.

**Additional conditions**

Letter Z is reserved for additional conditions drafted specially for a particular contract or included to comply with an employer's standard contractual requirements on such matters as confidentiality, discrimination, prevention of corruption and the like.

The promoters of the NEC recommend that additional conditions are kept to the absolute minimum and that they are drafted to match the style of the NEC using the same definitions and terminology. That being the case it is disappointing that the NEC does not have a greater range of standard secondary options. Some common contractual matters such as limitations of liability, assignment, patents, extraordinary traffic are not addressed in either the core clause or the secondary options and there will not be many NEC contracts where the employer does not find it necessary to include additional conditions for these or his own special needs.

Not all employers will have the ability or the inclination to draft their additional conditions in the unique style of the NEC – concerned, no doubt, in some cases that in doing so they might lose the true intention of an otherwise perfectly straightforward provision.

However, compatibility of drafting may not be a problem of any great magnitude. There is very little possibility that all the documents in an NEC contract will be written in matching style. And peculiar to the NEC is the way in which much of the detail found in other standard forms is excluded from the conditions of contract and required to be written into the works information. Details, for example, of taking over and performance tests, deliveries of plant and operating manuals.

It may, of course, be said that such detail should not be included in the works information since the works information is by definition in clause 11.2(5) information which either:

- specifies and describes the works, *or*
- states any constraints on how the contractor provides the works.

Taken literally the definition can be argued to exclude the detail of contractual obligations. But the problem is how to fix such detail into the contract if not in the works information. To put the detail into the additional conditions is even less attractive.

### Choice of secondary option clauses

The employer using either of main options A or B has full choice of the secondary option clauses and can include as few or as many as he wishes. There is not intended to be any duplication or inconsistency between the secondary options to restrict choice and nothing is obviously apparent.

If there is any possibility of overlap it may exist between Option M (limitation of the contractor's liability) and Option S (low performance damages) but that is a matter of detail discussed later in this chapter.

With main options other than A and B, however, there are some restrictions on which secondary options can be used. These are:

- Option K – multiple currencies
  Used only with main options A and B.

- Option N – price adjustment for inflation
  Not used with main options E and F.
- Option P – retention
  Not used with main option F.
- Option V – CDM Regulations
  Used only in the UK.

**Status of secondary option clauses**

The NEC, as discussed in Chapter 4, does not define which documents constitute the contract. Consequently it does not attempt to set any order of precedence for the various documents forming the contract. It leaves any ambiguities and inconsistencies to be resolved by the project manager under clause 17.1. This lack of any defined order of precedence together with the unusual 'pick and mix' arrangement of the clauses of the NEC may have some unintended effects.

One to note is that the usual rule of construction – the particular taking precedence over the general – is unlikely to apply as between core clauses and secondary clauses. So whereas in a traditional contract special conditions of contract take precedence over standard conditions in the event of ambiguity or inconsistency, the position in the NEC appears to be that special conditions included as additional clauses under secondary option Z have no precedence over the core clauses.

To overcome this, employers who see it as a problem should ideally include a clear statement identifying precedence in the contract. An alternative, but perhaps less certain method, would be to keep selected special conditions which are required to have precedence outside the scheme of secondary option clauses.

## 3.2 Option G – performance bond

Clause G1.1 requires the contractor to give a performance bond:

- for the amount stated in the contract data, *and*
- in the form set out in the works information.

The bond has to be provided by a bank or insurer which the project manager has accepted.

**Details of the bond**

Note that the contract data states only the amount of the bond, and the detailed form of the bond is to be set out in the works information. The difficulties that this may cause should not be underestimated.

Presently there is no model form of bond included in the NEC document pack and the model forms produced for use with other standard forms will not readily apply to the NEC because of differences in terminology. Special bonds will need to be drafted. This is most certainly not a task for amateurs. The drafting of bonds is a highly specialised business and the legal construction of bonds can perplex even the best lawyers. See, for example, the House of Lords decision in the case of *Trafalgar House* v. *General Surety & Guarantee Co.* (1995) on the much used standard ICE bond.

**Type of bond**

Bonds differ considerably in their drafting and in the conditions under which they can be called in for payment. At one end of the scale there are 'on-demand' bonds which can be called in without proof of default or proof of loss; at the other end of the scale there are 'performance' or 'conditional' bonds which can only be called in with certification of default and proof of loss.

The NEC refers in Option G to a performance bond. It is not clear whether this is intended to deliberately exclude the use of on-demand bonds with the NEC or whether it is simply general terminology which permits either type of bond.

**Acceptance of the bond**

Under clause G1.1 the project manager has discretionary power to accept or reject the bank or insurer proposed by the contractor as the provider of the bond. The only stated reason for not accepting the bank or the insurer is that its commercial position is not strong enough to carry the bond. Presumably it is intended that it is the project manager who should be the judge of this commercial position or at least nominally so. But the clause does not actually say this and the project manager will need to act with the greatest caution in rejecting any bank or insurer.

If the project manager is proved to be wrong in his assessment of the commercial position then the stated reason for non-acceptance is invalid and the minimum consequence is that compensation event 60.1(9) applies. Potentially worse is that the contractor fails to get another bond, that the employer then terminates under clause 95.2(R12), and that this is then held by an adjudicator or tribunal to be wrongful termination. It hardly needs to be said that the project manager's liability to the employer might then come under scrutiny.

**Provision of the bond**

The final sentence of clause G1.1 requires that if the bond is not given by the contract date it is given within four weeks thereof.

## 3.2 Option G – performance bond

The contract date is loosely defined in clause 11.2(3) as the date when the contract came into existence. Clearly the date needs to be positively fixed to give the provision in clause G1.1 effect. And since the only specified sanction in the NEC for non-provision of the bond is termination under clause 95.2 (R12) there is added need for certainty.

**Cost of the bond**

Unlike some other standard forms the NEC is silent on which party bears the cost of the bond, although ultimately, of course, whatever payment arrangements apply the cost should fall on the employer.

Unless the contract includes a method of measurement which states otherwise, the cost of the bond will be deemed to be included in the contract price.

For cost reimbursable contracts the cost of the bond is apparently to be included in the contractor's 'fee'.

## 3.3 Option H – parent company guarantee

Parent company guarantees give a measure of protection to the employer against a subsidiary contracting company's default and/or insolvency.

Because the financial strengths of subsidiary companies are not always reflected in their balance sheets, contracting companies which are subsidiary companies often put forward holding company accounts as evidence of stability. In such circumstances the employer may rightly decide that the security of a parent company guarantee is required in addition to (but sometimes as an alternative to) any performance bond which is specified. Clause H1.1 requires a contractor owned by a parent company to give a parent company guarantee:

- in the form set out in the works information
- within four weeks of the contract date.

Failure to provide the guarantee is a reason for termination under clause 95.2 (R12).

**Form of guarantee**

As with the performance bond the required form of guarantee is to be set out in the works information.

Drafters of the guarantee form should note a potential technical defect in the wording of clause H1.1, in that it refers to the guarantee being given by the company which owns the contractor. This is not necessarily the holding company (the ultimate parent company) within the terms of the Companies Act. Strictly, all that is required under clause H1.1 is a guarantee from a company owning the majority of the contracting company's shares.

## 3.4 Option J – advanced payment to the contractor

Advanced payments to contractors as intended by Option J are payments made as a matter of policy or trade custom. They have nothing to do with advanced payments which the contractor may obtain by front loading his activity schedule or bill of quantities.

A common reason for formal advanced payments is that the employer can secure funding at cheaper rates than the contractor; another is that the contractor has heavy early expenditure in procuring expensive plant and materials. Such payments are not uncommon in process and plant industries but are less so in construction.

### The amount

When Option J of the NEC applies, the employer's obligation is to pay the amount of advanced payment stated in the contract data (clause J1.1). Note that the clause is silent as to VAT. The point needs to be clarified in the contract data by writing inclusive or exclusive of any VAT which may be payable.

### Payment

Clause J1.2 requires the advanced payment to be made either:

- within four weeks of the contract date, *or*
- within four weeks of receipt by the employer of any advanced payment bond which is required, whichever is the later.

Delay by the employer in making payment is stated in the last sentence of clause J1.2 to be a compensation event.

### Security for advanced payment

When Option J1 is used the employer should indicate in the contract data whether or not a bond is required as security for the advanced payment.

Under clause J1.2 the bond is to be for the amount of the advanced payment and in the form set out in the works information. The bond is to be issued by a bank or insurer accepted by the project manager.

As with the performance bond, a reason for not accepting a bank or insurer is that its commercial position is not strong enough to carry the advanced payment bond. The potential consequences of rejection are similar for both bonds – see the comment above.

**Repayment**

Clause J1.3 requires any advanced payment to be repaid in instalments as stated in the contract data. The contract data deals with this by requiring two entries:

- the first stating when instalments are to commence, by reference to weeks after the contract date
- the second stating whether the instalments are amounts or a percentage of payments due (presumably to the contractor).

Where repayments are stated as amounts it does not automatically follow from clause J1.3 that repayment should only be by way of deductions from interim payments. That may be the broad intention of the scheme but the assessment procedures of Section 5 of the NEC contemplate the possibility that interim payments may be due from the contractor to the employer.

## 3.5 Option K – multiple currencies

The intention of Option K, when it is used, is to transfer the risk of exchange rate changes from the contractor to the employer. Its application is to firm price contracts rather than to cost reimbursable contracts and the NEC states that Option K should only be used with main options A and B.

Normally, Option K will apply only to parts of the works and these have to be specified in the contract data together with the applicable currency and the maximum payment to be made in that currency. Also to be stated in the contract data is the publication to be used as the source of relevant exchange rates.

## 3.6 Option L – sectional completion

Unless Option L is used the contractor's obligations to pay liquidated damages for late completion will apply only to the whole of the works.

Consequently, Option L is one of the more important secondary options for the employer to consider when putting together the contract.

In traditional contracts the problem frequently arises that employers intend partial completion dates to be contractually binding on the contractor but, although they clearly identify the parts, they state liquidated damages only for late completion of the whole of the works. The Courts, however, will not then award either liquidated or unliquidated damages for late completion of the parts. See, for example, the case of *Turner* v. *Mathind* (1986).

The same situation will normally arise under the NEC unless it is stated in the contract data that Options L and R apply and the description, completion date, and delay damages for each section are given.

There is, of course, the possibility under the NEC, because of the lack of precedence of documents and because the project manager is required to resolve ambiguities and discrepancies between documents, that, in the event of some documents showing sectional completion when the contract data does not, the project manager could give instructions imposing sectional completion requirements. The contractor would have no liability for liquidated damages for late completion of such sections but he would be entitled to a compensation event in respect of the instructions.

**Clause L1.1 – sectional completion**

In many standard forms of contract the provisions for sectional completion are lengthy and complicated, but the NEC uses the simple device of stating in clause L1.1 that each reference and clause relevant to the works, completion and the completion date applies, as the case may be, to either the whole of the works or to any section. Where, however, the phrase 'the whole of the works' is used in the conditions of contract, that phrase is not to be taken as applying to sections. In short, when Option L is used, the NEC distinguishes between 'the works' and 'the whole of the works'.

Note, however, that to make the NEC arrangement work, the total of the sections should not comprise the whole of the works, and that delay damages should be stated for both sections and the whole of the works.

For the possibility of having a combination of liquidated and unliquidated damages for delay under the NEC, see the section later in this chapter on Option R.

## 3.7 Option M – limitation of the contractor's liability for his design to reasonable skill and care

Option M is the shortest of the secondary option clauses but it is the one which has attracted the most comment from lawyers.

## 3.7 Option M

The option states (clause M1.1) that the contractor is not liable for defects in the works due to his design so far as he proves that he used reasonable skill and care to ensure that it complied with the works information. It may well be that the draftsmen of the NEC intended nothing more in this than that when the option is used there should be no implied term in the contract that the contractor's liability for his design should be on a fitness for purpose basis.

The probability of such an implied term in design and build contracts was suggested by the House of Lords in the case of *IBA* v. *EMI* (1980). And because it imposes a standard of liability which is potentially higher than the standard of liability carried by professional designers (skill and care), contractors have argued with some success in relation to many standard forms of contract that a clause should be included limiting their liability for their design to skill and care.

The problem with clause M1.1 of the NEC, however, is that on its particular wording it is open to various interpretations, some of which far from limiting the contractor's liability might actually increase it.

### Peculiarities of clause M1.1

Full analysis of the legal effects of the peculiarities of clause M1.1 is beyond the scope of this book but in short the points to note are:

- The clause applies to defects in the works due to the contractor's design rather than to the design itself, with the possibility that the limitation could apply to defects in works not designed by the contractor.

- The limitation of liability in the clause applies only to 'Defects' within the meaning of the defined term.

- The clause reverses the usual burden of proof applying to negligence so that the burden of proof is put on the contractor to show that he used reasonable skill and care. Consider the difficulties of this for a contractor where a specialist subcontractor has been involved as the designer and is no longer in business.

- The obligation to use skill and care applies only to compliance with the works information – and, for contractor's design, much of this may have been provided by the contractor.

- The effects of the application of the clause before completion are uncertain.

### An alternative approach

Employers who intend nothing more from Option M than a simple change in the contractor's liability for his design would do well to look at the approach of the standard building design and build contract JCT 1981. In effect that contract says in clause 2.5.1 that the contractor's liability for his design is the same as that of a professional designer.

## 3.8 Option N – price adjustment for inflation

Price fluctuation clauses are commonly included in standard forms of contract as an optional extra and that is the policy of the NEC.

Option N is a conventional formula/index based fluctuation clause which allows adjustments to the contract price for inflation. It is used only with main options A, B, C and D. Its use is unnecessary with main options E and F which are fully cost reimbursable. When Option N is used the particulars governing the application of the formula and the indices must be included in the contract data.

## 3.9 Option P – retention

Provisions entitling the employer to retain a percentage of amounts due to the contractor until the works are completed and any defects period has expired are standard in most construction, process and plant contracts. The NEC, however, makes this a secondary option rather than a core clause and employers will have to be careful that it is not inadvertently omitted from the contract.

**Clause P1.1 – deduction of retention**

Clause P1.1 deals with the deduction of retention. The first point to note is that the deduction of retention is not intended to commence until the valuations have reached a 'retention free amount'. This is intended to assist the contractor's cash flow in the early stages of the contract. The retention free amount is to be entered in the contract data. If no such amount is entered, or nil is entered, the deduction of retention will commence from the first valuation.

The amount of retention is determined by the 'retention percentage' which is entered in the contract data. If this is left blank the employer will have no entitlement to retention. The retention percentage is applied only to the excess above the retention free amount and not to the whole of any valuation.

One aspect of clause P1.1 which may cause some concern to contractors is that retention is apparently held against sums valued for compensation events. This seems to follow from the definition of the price for work done to date. However, it is hardly equitable that the employer should be entitled to retention on an amount payable to the contractor in respect of any compensation event which is a breach of contract by the employer. This is not permitted in many standard forms.

**Clause P1.2 – release of retention**

Clause P1.2 deals with the release of retention. The approach is conventional. Half the retention is released on completion and the remainder on the issue of the defects certificate.

**Trust status**

There is nothing in Option P stating that retention is held in trust or requiring the employer to hold retention in a separate bank account. However, see the case of *Wates Construction* v. *Franthom Property* (1991) on the possibility of implied terms that the employer is a trustee and has a duty to safeguard the interests of the beneficiaries (the contractor and subcontractors).

Consider also the effect of the obligation in clause 10.1 for the parties to act in a spirit of mutual trust and co-operation. Could this extend to an obligation to hold retention money in a trust fund?

## 3.10 Option Q – bonus for early completion

Provisions for payment to the contractor of a bonus for early completion are not common in standard forms but ad hoc arrangements for such payments are not unusual. The NEC sensibly includes the bonus provisions as a secondary option.

**Clause Q1.1 – bonus for early completion**

The drafting of clause Q1.1 is comparatively straightforward in that it provides for the contractor to be paid a bonus:

- calculated at the rate stated in the contract data
- for each day from the earlier of
  - completion of the works
  - take over of the works
  until the completion date.

Note however that the only figure to be entered in the contract data is the rate per day for 'the whole of the works'. Therefore, although clause Q1.1 on its wording might apply to sections, it will probably not do so unless the contract data is extended to include additional figures for sections.

**Apportionment**

It may, of course, be argued that the wording of clause Q1.1 implies that there should be apportionment of the daily rate of bonus for the whole of the works when there is early completion or take over of any part of the works. The basis of such an argument is that since the clause states that the bonus becomes payable from the earlier of 'completion' or 'the date on which the employer takes over the works', and the contract provides for take over of parts of the works, then some bonus should be paid when parts are taken over before completion.

For example, a situation might arise when 90% of the works are taken over early and put to use by the employer, but completion of the whole is not certified by the project manager until the due completion date. How then would the contractor have any entitlement to a bonus without apportionment? The strict legal approach may be that there is no provision for apportionment and the contractor has no entitlement to a partial bonus, but this is hardly in the spirit of the contract.

**Effects of delays**

Although the NEC may be subject to argument on apportionment in its bonus provisions, it does appear to have eliminated one of the commonest causes of argument found with similar provisions in other contracts. That is the question of whether delays for which the employer is responsible, or any delays which give entitlement to extension of time, should be taken into account in calculating the bonus. In some contracts the completion date is fixed for the purposes of calculating the bonus but in others it is not clear how delays should be treated or whether extensions of time apply to bonuses.

In the NEC, because the compensation event procedure moves the completion date (whether or not an extension of time is required to avoid delay damages), the contractor's bonus entitlement is protected against any delay which is a compensation event.

In the event that acceleration is considered under clause 36 of the NEC, the parties will have to give some thought to how that relates to Option Q.

## 3.11 Option R – delay damages

A contractor who fails to complete by the due date is liable to the employer for damages for breach of contract. Such damages may be either specified in the contract (and are then usually known as liquidated damages) or they may be left to be determined after the breach of contract as general damages.

### Liquidated damages

When damages are liquidated they can be seen either as providing compensation for the employer in lieu of general damages, or they can be seen as limiting the contractor's liability for his breach of contract. They serve as both and are regarded in law as an exclusive and exhaustive remedy. See the case of *Temloc* v. *Errill* (1987).

To be enforceable (and not liable to challenge as penalties) any sum specified as liquidated damages must be a genuine pre-estimate of the employer's loss or a lesser sum. And, because the courts have traditionally taken a strict approach to the construction of provisions for liquidated damages, to be effective such provisions must be clear and unambiguous.

### Delay damages in the NEC

The core clauses of the NEC are silent on delay damages so unless secondary option R for delay damages is included, the legal position is almost certainly that:

- the employer retains his common law rights and can sue for the damages he can prove he has suffered as a result of the contractor failing to complete by the due date, *and*
- the contractor is liable for the full amount of those damages unless the contract contains some general limitation of his liability.

Option R is not named a liquidated damages clause but it is clearly intended to operate as such. It requires a rate for damages to be entered in the contract data and the presumption must be that the rate conforms with the rules for liquidated damages. If not, Option R is unenforceable and pointless.

An awkward legal question could arise in the event of Option R being listed in the contract data as applicable to the contract, but no rate being set in the contract data for the delay damages. The question might then be asked: does the inclusion of Option R act, in itself, as an exclusion of the employer's common law right to general damages? Or, to put it another way, would an employer by including Option R but failing to state a rate for delay damages, forgo his right to any delay damages, liquidated or otherwise?

In the case of *Temloc* v. *Errill* mentioned above, the employer, under a JCT contract, wrote £nil as the rate for liquidated damages. The Court of Appeal held that the contractual provision for liquidated damages remained valid and therefore the employer had lost his common law remedy of general damages. The case is arguably not applicable to a blank rate entry, as opposed to a £nil rate entry, but against that the express inclusion of the delay damages option clause might be persuasive of the parties' intention that general damages should be excluded. Employers should note the point and ensure that the difficulty does not arise.

### Application to sections

Clause R1.1 of the NEC does not expressly mention sections, but the intention that Option R should apply to sections, if so desired, is evident from the layout of the contract data sheet. This has spaces for the inclusion of rates for delay damages for sections, as well as a space for the rate for the whole of the works.

The application of the delay damages provisions in clause R1.1 to sections relies entirely on the effectiveness of Option L (sectional completion) in giving the phrases 'the works, completion and completion date', both singular and plural meanings.

It remains to be seen what the courts will make of this. They may take the view that the secondary options bolt independently onto the core clauses and are not to be interpreted as relating to one another. If that happens the provisions for liquidated damages for sections in the NEC will fail. Until the point is resolved employers concerned over the matter might consider expanding the wording of Option R with some express reference to sections.

### Proportioning down of delay damages

It is, of course, going against the recommendations of the promoters of the NEC to change the wording of core or option clauses. But, for delay damages, many employers will see that as necessary quite apart from the reason above on sectional completions.

The problem is that the NEC, most unusually, does not contain any provisions for proportioning down the rate (or rates) of liquidated damages when parts of the works are taken over or certified as complete before the whole. The need for proportioning down clauses has long been recognised and they are found in all other well used standard forms. They

## 3.11 Option R – delay damages

protect the stated rates of liquidated damages from being declared penalties – the argument being that once part of the works is taken over or certified complete, the stated rates are no longer a genuine pre-estimate of the employer's loss for the remainder. See, amongst others, the case of *Bramall & Ogden* v. *Sheffield City Council* (1983).

The case for having a proportioning down clause in the NEC is particularly strong because the NEC expressly provides for take over of parts of the works (clause 35.4). Employers are recommended therefore to look at the wording of other standard forms and to devise a suitable addition to Option R. The proviso in clause 47.1 of the ICE 6th edition is readily adaptable.

### Combination of delay damages

The prospect was mentioned above in comment on Option L (sectional completion) that it might be possible to combine within the NEC both liquidated damages for the whole of the works and general (unliquidated) damages for sections, or vice versa. In principle there appears to be nothing against this providing there is no double recovery of damages. In *Turner* v. *Mathind* (1986) Lord Justice Parker expressed quite firmly the view that liquidated damages for the whole of the works should not necessarily exclude general damages for sections.

The option structure of the NEC seems to lend itself to this arrangement and it could arguably be achieved by including both Options L and R in the contract and by stating applicable, or not applicable, as appropriate in the rates entries of the contract data. But to ensure certainty, employers who wish to combine liquidated and general damages in an NEC contract are advised to take legal advice on how the entries should be made and what changes (if any) should be made to Options L and R.

### Clause R1.1 – payment of delay damages

The key points of clause R1.1 are:

- the contractor pays delay damages
- at the rate stated in the contract data
- from the completion date
- for each day
- until the earlier of:
  - completion, 'and'
  - the date of take over

Note firstly a small semantic point – the use of the word 'and' where 'or' would seem more appropriate. This occurs frequently throughout the NEC. Sometimes it is of no obvious consequence, but sometimes it is a point of major concern. See for example the comment in Chapter 9, section 3, on clause 40.1.

More importantly, however, note the absence in clause R1.1 of any of the usual conditions precedent to the deduction of delay damages, e.g.:

- certification of failure to complete on time
- certification that no further extensions of time due
- notification of intention to deduct.

It may well be that the NEC omits reference to these customary formalities in the interests of simplicity. The consequences, however, may be anything but simple and they are potentially adversarial. What may have been lost by the terse wording of clause R1.1 is the employer's discretion whether or not to deduct damages to which he is entitled.

The scheme appears to be that under clause 50.2 the project manager assesses the amount due taking into account any amounts 'to be paid by or retained from the contractor'. The employer then pays the amount due. Hence the employer's loss of discretion. But what of the position if the project manager fails to deduct for damages in his assessment and the damages are not then paid when they become due? Can it then be argued that the employer has waived his right to damages?

**Completion and take over**

Note that the project manager decides the date of completion under clause 30.2 and that completion is defined in clause 11.2(13).

Clause 35.3 defines the meaning of take over and clause 35.4 requires the project manager to certify the date on which the employer takes over any part of the works.

**Clause R1.2 – repayment of damages**

Clause R1.2 deals with repayment by the employer of delay damages when the completion date is changed to a later date. The clause provides that the employer repays the 'overpayment of damages' with interest. The rate of interest is not stated but presumably the interest rate inserted in the contract data and referred to in clause 51.5 is intended to apply.

The phrase 'overpayment of damages' suggests only partial repayment but it is unlikely to be so limited in its application. For comment on how changes in the completion date are assessed and made, see Chapter 11, section 1.

The final words of clause R1.2 – 'and the date of repayment is an assessment date' – are something of an enigma. The grammar of the sentence is difficult to follow. However, what is probably meant is that interest runs not to the repayment date itself but only to the date of the project manager's assessment of the repayment.

## 3.12 Option S – low performance damages

Performance testing is routine in process and plant contracts and it is occasionally found in construction contracts. The NEC, as mentioned in Chapter 1, is intended to be suitable for process and plant installations as well as for construction works. However, unlike conventional process and plant contracts, the NEC does not concern itself with procedures for performance testing. It leaves those to be specified with the detail of any performance tests (and other tests) in the works information.

The only specific mention of performance in the NEC is in secondary option S – low performance damages. However, it should not be assumed from this that unless Option S is included in the contract the employer has no remedy for low performance. It may well be that the Section 4 core clauses on testing and defects, which are written in the most general terms, can have application to performance specifications.

That does raise an important question as to the purpose of Option S when it is included. Is it to be seen as providing an additional remedy to the employer for low performance, or is it to be seen as an alternative and exclusive remedy? The answer, in any particular case, may depend upon how the low performance damages are stated in the contract data. For example, if the damages are clearly only applicable within a particular range of defaults, then Option S may be limited in its effect. But otherwise a significant, although perhaps unintended, effect of Option S may be to limit the contractor's liability rather than to fully compensate the employer. The legal reasons for this are explained above in the comments on liquidated damages in Option R (delay damages). The contractual reasons are that the NEC provides a number of alternative remedies as discussed below.

### Performance testing

Performance tests are tests made to establish whether the works operate as specified under working conditions. Such tests can readily be distinguished from other tests because:

- they are usually made after take over when the works are in use, *and*
- they are usually carried out by the employer, with the contractor observing or supervising.

## 3.12 Option S – low performance damages

Various possible remedies for the failure of performance tests exist according to the seriousness of the failure and the express terms of the contract:

- modification of the malfunctioning works
- payment of liquidated damages for low performance
- acceptance of the works subject to a reduction in the contract price
- rejection of the works on the grounds of unacceptably low performance.

None of these is wholly without its difficulties, particularly in relation to timing and compatibility with other contractual provisions. The problem essentially is that the contractor usually has, before commencement of performance tests, a take over certificate which signifies that he has fulfilled his obligation to construct and complete the works.

**Remedies under the NEC**

Under the NEC the possible remedies for low performance are:

- clause 43.1 – the contractor is required to correct defects
- clause 44.2 – the contract price is reduced in acceptance of a defect which is not corrected
- clause 45.1 – the contractor is liable for the cost of having defects corrected by others
- Option S – low performance damages.

Having regard to the variety of these remedies and to the implications of limiting liability by the use of liquidated damages, it is recommended that users of the NEC take legal advice before including Option S in their contract.

A point worth noting is that because the NEC has no express provision for rejection of the works if performance is wholly unacceptable (such as clause 35.8(c) of MF/1), an employer under an NEC contract may have difficulty in establishing any legal right to reject. This is because it is doubtful if common law rights or statutory rights to reject apply to works incorporated into the employer's property.

**Clause S1.1 – low performance damages**

The essentials of clause S1.1 are:

- a defect must be included in the defects certificate

- the defect must show low performance against a performance level stated in the contract data
- the contractor pays the amount of low performance damages stated in the contract data.

Firstly, note that the clause applies only to a defect within the scope of the defined term. That means that the defect will have to be measurable by way of some criteria in the works information or in the contractor's design. Secondly, however, note that low performance is to be measured against a performance level stated in the contract data. That means there will have to be a careful link between the criteria in the works information and the statements in the contract data.

In practice it will rarely be possible to state the low performance damages as simply as the contract data sheet seems to indicate. Amongst the matters commonly subject to performance tests are:

- ability of the works to achieve quoted efficiency
- power consumption
- cost of operating and maintaining the works
- product quantity and quality
- consumption of chemicals
- quantity and quality of effluents
- volume of waste products
- pollution and noise control.

Each requires its own parameters and it is not unusual for statements on low performance damages to run to many pages of print.

As to the mechanism for the payment of low performance damages, clause S1.1 is silent. Possibly the Section 5 rules on payment are intended to apply. But note that if that is the case, clause 50.1 is restrictive of when assessments are made and the only relevant assessment for low performance damages may be that made four weeks after the issue of the defects certificate.

## 3.13 Option T – changes in the law

The purpose of Option T is to place with the employer the risks of contract costs and completion times being affected by changes in the law. Option T does this by making changes in the law compensation events. The same effect could probably be achieved by making changes in the law the employer's risk under clause 80.1, so that they then come within the scope of compensation event 60.1(14).

Without Option T the amount of risk carried by each of the parties on changes in the law depends upon which of the main options is used.

## 3.13 Option T – changes in the law

Generally the contractor takes the risks on time under all the main options, but the risks on price follow the usual rules for the main options.

**Clause T1.1 – changes in the law**

Clause T1.1 operates only when there is a change in the law of the country in which the site is located and the change occurs after the contract date, i.e. the date on which the contract was made. This is slightly different from some other contracts where changes in the law which occur after the return of tenders are taken into account.

**Procedures**

Clause T1.1 does not rely expressly on the procedures for the notification and assessment of compensation events set out in Section 6 of the NEC. It states some procedures of its own:

- the project manager may notify the contractor of a compensation event
- the project manager may instruct the contractor to submit quotations
- the prices are reduced for changes which reduce total actual cost.

It is not clear if these stated procedures are meant as supplements to the Section 6 procedures or as partial replacements. But for contractors the key question is probably, does the two week notice rule of clause 61.3 apply? The safe answer is to assume that it does.

However, one curious and obviously unintended effect of applying the Section 6 procedures rigorously to clause T1.1 is that the contractor is obliged under clause 61.3 to give notice of all changes in the law of the country occurring after the contract date. It is then for the project manager to decide under clause 61.4 whether or not the changes have any effect on cost or completion.

## 3.14 Option U – The Construction (Design and Management) Regulations 1994

Option U on the effects of the CDM Regulations is applicable only to contracts in the UK. Is it, however, an essential option for all NEC contracts in the UK? The words in brackets which follow the heading of Option U ('to be used for contracts in UK') suggest that it is. But that is not consistent with the statement in the NEC Schedule of Options:

> 'The following secondary options should then be considered. It is not necessary to use any of them.'

Looking at what Option U actually sets out to achieve – the placement of certain risks on the employer – there is no obvious reason why Option U should be regarded as different from the other secondary options in the matter of choice.

**Clause U1.1 – the CDM Regulations**

Compared with the wording of CDM clauses in other construction contracts the wording of clause U1.1 is neat and economical. It states simply that:

- a delay to the work, *or*
- additional works, *or*
- changed work
- caused by the application of the Regulations
- is a compensation event
- if an experienced contractor could not have foreseen it.

At the present time the full impact of the application of the CDM Regulations has yet to be felt and experience with claims arising from their application is limited. Concern has been expressed in various quarters at the costs an over zealous planning supervisor could cause and that the certainty of price of lump sum design and build contracts may have been eroded.

Employers considering the use of Option U will, no doubt, take into account whether or not they wish to take on additional financial risk. And the point should, perhaps, be made that Option U may be no more than additional risk so far as the application of the CDM Regulations is concerned. Many potential claims on costs and time arising from the Regulations could well be covered by the standard compensation events.

It has even been said that on strict interpretation Option U can be taken as restricting the contractor's entitlements, not enhancing them. That is because clause U1.1 makes the contractor's entitlement subject to a foreseeability test. Such a test does not apply to other compensation events which might otherwise cover the relevant delays or changes.

## 3.15 Option V – trust fund

The NEC is the first standard form of contract to make provision for a trust fund. The merits of such a fund have been much debated and in simple terms the principal argument in favour of the fund is that all the participants in the project benefit when there is protection against financial instability.

There are, of course, ways of obtaining protection other than by trust funds. For example, bonds – which are widely used to give protection party to party in the downward direction of the procurement chain – can be used in reverse to give upward protection.

But Option V of the NEC intends a far wider scheme of protection than can be obtained by any party to party bonding arrangements. It intends that the employer should set up a trust fund with all the parties in the supply chain as potential beneficiaries. So under Option V the trust fund is not simply protection to the contractor against the employer's insolvency. It is of far wider application. In effect, up to the level of the fund, the employer acts as guarantor to all subcontractors and suppliers against insolvency anywhere in the chain. And, in the event that there is payment out of the trust fund, the employer tops up the fund to maintain it at its initial value – a process to be repeated as and when necessary.

**The initial value**

Clause V1.1(3) states that the initial value of the trust fund is 1.5 times the contract price divided by the number of months allowed for completion. For example, for a £5 million contract to be completed in ten months the initial value of the trust fund is £750,000.

Under clause V2.2 the trust fund is to be established within one week of the contract date. Any replenishment which is necessary is made within two weeks (clause V3.1(4)).

**The trustees**

The trust fund is administered by trustees in accordance with a trust deed (clause V3.1). The trustees have discretion to decide the amount and timing of any payment from the fund (clause V3.1(3)) and they are entitled to their fees and expenses (clause V3.1(6)). In exercising their discretion the trustees may have a mass of conflicting claims and counter-claims to consider and their task may be anything but easy.

**The response to Option V**

The initial response to Option V has been one of surprise at the extent of the potential liability it imposes on the employer. It appears to go far beyond what was expected or what was intended.

It is thought unlikely that any professional advisor will recommend to an employer that Option V be included as it stands with any of the main options of the NEC. Accordingly there is little point in continuing here with a detailed analysis of the lengthy text of Option V.

# Chapter 4

# The contract documents

## 4.1 Introduction

Written contracts are made to bring certainty to the intentions of the parties. Fundamental to that is certainty as to which documents form the contract. The NEC leaves this certainty to be provided by the parties in so far as the NEC, unlike other standard forms, does not explicitly define the contract or list the documents forming the contract.

It may have been thought that such a definition or list would be restrictive to the use of the NEC; or, put another way, that no single definition or list would cover the full range of possible uses of the NEC. But, whatever the reason, users of the NEC need to be alert to the fact that the NEC is not only different from other contracts in its style and philosophy but is also significantly different in its approach to standard contract documentation.

Particular points to note are:

- the NEC does not contain any standard
  - form of tender
  - articles of agreement
  - recitals
  - form of agreement
  - form of performance bond
  - forms of guarantee
- the NEC is silent on how post tender negotiations and correspondence should be given contractual effect
- the NEC has no stated order of precedence of documents
- the NEC does not say whether or not headings and side notes in the conditions of contract are to be considered in construction of the contract
- the NEC has no place to state (and record) the very important contract date.

These are the peculiarities of the NEC by way of omissions and some, if not all of them, will have to be addressed by compilers of NEC contracts.

An equally important, but perhaps less obvious, task which falls on compilers of NEC contracts is that of ensuring that the works information details all those matters necessary to the operation of the conditions of contract. The point is dealt with in some detail later in this chapter and elsewhere in this book. But it cannot be said too often that the NEC relies on detail in the works information to specify the contractor's obligations to an extent which has no parallel in any other standard form of contract.

The overall effect of the differences between the NEC and other standard forms of contract is that putting together a competent set of NEC documents is a more formidable task than usual. A secondary effect is that it exposes compilers of NEC contracts to greater risk of error and liability in negligence.

## 4.2 Sequence of documentation

In a typical NEC contract the following sequence of documentation will probably apply, although not all the documents listed here will necessarily become contract documents.

### Documents sent to tenderers

- instructions to tenderers
- contract data – part one
- works information including
  - scope of works
  - drawings
  - specifications
  - other details as appropriate
- site information
- bill of quantities (if appropriate).

### Documents submitted with tenders

- form of tender
- contract data – part two
- works information (from the contractor)
- contractor's price, including (as appropriate)
  - activity schedule
  - bill of quantities
  - cost component data
- programme (if required).

### Documents making the contract

- letter of acceptance
- form of agreement.

*Documents submitted post award*

- performance bond
- parent company guarantee
- programme.

**Instructions to tenderers**

The NEC is silent on the status of instructions to tenderers and it is a matter of policy for individual employers whether or not they are to become contract documents. For the avoidance of doubt a clear statement should be made at the head of instructions to tenderers stating whether or not they are to form part of the contract.

## 4.3 Essential contract documents

As an absolute minimum the essential documents of an NEC contract are:

- contract data – part one – provided by the employer
- the contractor's pricing document
- conditions of contract with
  - main option stated
  - secondary options stated
- works information with input as appropriate from
  - the employer
  - the contractor
- site information – provided by the employer
- contract data – part two – provided by the contractor
- form of tender
- letter of acceptance.

**Incorporation of the NEC conditions of contract**

An important question is whether the NEC conditions of contract (the standard core clauses and secondary options) can be incorporated into particular contracts by reference or whether it is necessary in each case for a printed copy of the conditions of contract to be included in the bundle of documents and signed by the parties.

The opening statement in part one of the contract data is apparently intended to incorporate the conditions of contract by reference and in many cases the parties will accept that as sufficient. There is, however, an important legal point to consider: whether it is possible to incorporate an arbitration agreement by reference, even when the arbitration agreement is itself contained in a set of standard conditions. See the Court of Appeal decision in *Aughton* v. *Kent* (1991) which appears to suggest that as a general rule the incorporation of arbitration agreements by reference is not effective. And note the more recent decision in *Ben Barrett & Son (Brickwork) Ltd* v. *Henry Boot Management Ltd* (1995) following the decision in *Aughton*.

This is matter on which the parties should take legal advice if they have any doubts. If the decision holds good for the NEC then the effect of incorporation by reference may be that much of Section 9 of the NEC could inadvertently be lost.

**Form of tender**

The NEC itself does not have a model form of tender although the second edition Guidance Notes do have a sample form of tender. The essentials of a form of tender are:

- that it should constitute an offer from one named party to another
- that it should describe the goods/services/works to be provided
- that it should state the price of the offer (or applicable pricing mechanism)
- that it should state the basis of the offer by reference to the tender documents.

For an NEC contract the offer itself can perhaps be put along the following lines:

> 'We offer to provide the Works as described above in accordance with Parts One and Two of the Contract Data for these Works for such sum as may be ascertained in accordance with the Conditions of Contract specified in the Contract Data'.

But again this is a matter on which legal advice should be taken.

**Form of agreement**

With some standard forms of contract (such as the ICE Conditions) completion of a form of agreement is necessary only when the parties intend to execute the contract as a deed, thereby extending the limitation period under English law from 6 to 12 years. With other standard forms, however, (such as the IChemE contracts) the form of agreement is the definitive document which is evidence of the contract and its contents. It is then an essential contract document.

With the NEC, because of the potential uncertainty over which documents are to be regarded as contract documents, there is a good case for saying that a properly drafted form of agreement should be used in all cases. Again, the second edition of the NEC Guidance Notes has a sample form of agreement.

## 4.4 Identified and defined terms

The NEC uses a system of identified and defined terms which clause 11.1 indicates to be as follows:

- identified terms are terms particular to the contract, such as names and details, which are stated for each contract in the contract data
- defined terms are terms, general to the NEC conditions of contract, which are given a particular meaning by definitions in clause 11 of the NEC.

The general scheme intended by clause 11.1 is that within the text of the NEC, identified terms are in italics and defined terms have capital initials. There are, however, some anomalies:

- not all terms with capital initials are defined terms; some such as project manager, supervisor, adjudicator, employer and contractor are in fact identified terms
- not all terms which are identified in the contract data are in italics; note, for example, works information and site information
- not all terms which are effectively defined are treated as defined terms; note in particular, 'compensation event'.

## 4.5 The contract date

Clause 11.2(3) defines the contract date as the date when 'this contract' came into existence.

Obviously in the case of the contract date the phrase 'this contract' refers to a particular contract and not to the NEC itself. However, in some clauses of the NEC the phrase 'this contract' does appear to be a reference to the general text of the NEC. In clause 29.1 (instructions) it is a matter of great importance whether the term 'this contract' is meant to be specific or general.

**References to the contract date**

The purposes of references in the NEC to the term 'contract date' are principally to set time running for the commencement of certain obligations and to establish base dates relevant to certain compensation events. The references are:

- clause 60.1(12) – compensation event for unforeseen conditions
- clause 60.4 – compensation event for remeasurement with bill of quantities
- clause 60.5 – compensation for delays caused by increased quantities with bill of quantities
- clause G1.1 – provision of performance bond
- clause H1.1 – provision of parent company guarantee
- clause J1.2 – making advanced payment
- clause T1.1 – changes in the law
- clause V1.1 – amount of trust fund
- clause V2.1 – establishment of trust fund
- see also clause 31.1 (provision of first programme) which links with the contract date via the contract data.

The NEC, however, does not use the contract date as the general date for the commencement of all obligations. The contract date, for example, has no contractual link with the completion date (which is set independently in the contract data). For the commencement of some obligations and functions the NEC uses other terms:

- the starting date
- the possession date.

*The starting date*
The starting date is not a defined term of the NEC. It is an identified term fixed in part one of the contract data by the employer. Its purpose is principally administrative.
Reference to the starting date is made in the following clauses:

- clause 20.4 – forecasts of total actual cost in Options C, D, E and F
- clause 32.2 – submission of revised programmes
- clause 50.1 – fixing assessment dates for payments
- clause 81.1 – contractor's risks
- clause 84.1 – insurance cover
- clause 85.1 – insurance policies of the contractor
- clause 87.1 – insurance policies of the employer.

*The possession date*
Like the starting date the possession date is not a defined term but an identified term fixed by the employer in the contract data. It relates to the commencement of work on site.

## 4.5 The contract date

References to the possession date in the NEC are as follows:

- clause 30.1 – starting work on site
- clause 33.1 – possession of the site
- clause 60.1(2) – compensation event for late possession.

## 4.6 Works information

Works information is at the heart of the NEC and unless it is competently prepared key parts of the contract will be of no application. Works information is defined in the first part of clause 11.2(5) as information which either:

- specifies and describes the works *or*
- states any constraints on how the contractor is to provide the works.

This definition suggests that the works information is little more than a specification, but to grasp the full extent of the importance of the works information see the schedule at the end of this chapter setting out the many references in the text of the NEC to the works information. The count runs to 40 references.

Not all of these references require the entry of detail in the works information but nevertheless, from examination of those that do, it is clear that the works information is much more than the specification for the works.

It is probably of no contractual significance that the works information goes beyond the apparent scope of its definition. But, of course, to any compiler of an NEC contract and to the parties using the contract it is a matter of great significance that all necessary information and detail is included.

These are the matters which should be considered:

- clause 11.2(13) – work to be done by the completion date
- clause 18.1 – health and safety requirements
- clause 20.1 – the works to be provided
- clause 20.2 – subcontracting requirements
- clause 21.1 – extent of contractor's design
- clause 21.2 – particulars of contractor's design to be submitted for acceptance
- clause 22.1 – use of the contractor's design
- clause 25.1 – sharing of the working areas
- clause 31.2 – works by the employer and others
- clause 31.2 – information to be shown on the contractor's programme

- clause 33.2 – provision of facilities and services
- clause 35.3 – use of part of the works before take over
- clause 40.1 – tests and inspections
- clause 40.2 – materials, facilities and samples for tests and inspections
- clause 52.2 – accounts and records
- clause 71.1 – marking of equipment, plant and materials
- clause 73.2 – title to materials from excavation and demolition
- clause G1.1 – performance bond
- clause H1.1 – parent company guarantee
- clause J1.1 – advanced payment.

**Identifying the works information**

Given the range of information which can constitute works information, how is the contractor to know what can properly be described as works information? The second part of clause 11.2(5) deals with this. The clause states that works information is either:

- in the documents which the contract data states it is in, *or*
- in an instruction given in accordance with the contract.

In part one of the contract data the employer should state which documents his works information is in. And in part two of the contract data the contractor should state which documents the works information for any contractor's design is in.

This rigid scheme for identifying the works information has its obvious drawbacks. A document may be listed as a contract document but then if an omission is made in the listing in the contract data for works information in respect of that document, its effect may be nullified. The project manager could rectify such a situation under clause 17.1 (ambiguities and inconsistencies) or clause 14.3 (instructions), but compensation event procedures would then apply.

**Instructions as works information**

As noted above, clause 11.2(5) allows works information to be in an instruction given in accordance with the contract.

Clause 14.3 allows the project manager to give an instruction which changes the works information. So also does clause 19.1 (illegal or impossible requirements) and clause 44.2 (accepting defects). However, it is not clear whether these clauses apply only to 'changes' in the works information or whether they apply also to additional information – which then creates new obligations.

## 4.6 Works information

For example, if there is nothing in the works information requiring the contractor to design parts of the works, is the project manager empowered to give instructions which impose design obligations on the contractor? Would such an instruction be a 'change' in the works information or a completely new category of works information? Would such an instruction be an instruction given in accordance with the contract?

There are no certain answers to these questions. But they are important because the NEC has none of the usual provisions fixing the scope and limitations of variations which may be ordered under the contract. All that the NEC has is:

- clause 20.1 requiring the contractor to provide the works in accordance with the works information
- clause 11.2(5) stating that works information is information which is in an instruction given in accordance with the contract
- clause 14.3 empowering the project manager to give an instruction which changes the works information.

Clearly some practical, if not contractual, limitations have to apply otherwise every NEC contract is completely open-ended as far as the contractor's obligations are concerned. A contract to build a school could theoretically be turned into a contract to build a hospital. No one, of course, would attempt to argue that such an extreme change was valid, but that will not remove the scope for argument on lesser changes.

Perhaps in the course of time the courts will lay down some guidelines on how this aspect of the NEC is to be interpreted. Until then, those with a problem should focus on the word 'information' in clause 11.2(5). Perhaps it is possible to argue that it is simply 'information' in an instruction which constitutes works information and that a bald instruction is not information. See also the comment in Chapter 6, section 3 on the definition in clause 11.2(4) of 'to Provide the Works'.

### Works information provided by the contractor

The only works information which the contractor is entitled or required to provide is that relating to his design. In many cases this may be minimal or non-existent. But in other cases, where the contract is fully contractor designed, most of the technical input to the works information may be provided by the contractor. This will certainly be the case when the employer states his requirements in performance terms.

Only clause 11.2(15) – definition of defect, and clause 63.7 – assessment of compensation event for ambiguity – expressly recognise that works information provided by the contractor should be treated differently from works information provided by the employer. But, as indicated in comment on various clauses of the NEC later in this book, it may be an oversimplification of the contractual position that otherwise the NEC treats works information provided by the contractor and works information provided by the employer alike.

**Inadequate works information**

There is no express obligation in the NEC on the employer (through the project manager) to issue additional works information beyond that in the documents identified in the contract data. Consequently it is not absolutely clear how the position on inadequate information is to be resolved.

One contractual solution would be the calling of an early warning meeting under clause 16.2, followed by the project manager giving instructions under clause 14.3 (instructions) or clause 19.1 (impossibility). But in the event of this solution failing, the contractor might have to rely on there being an implied term in the contract that the employer is obliged to provide all information necessary for the works to be completed.

## 4.7 Site information

Site information is defined in clause 11.2(6) as information which:

- describes the site and its surroundings, *and*
- is in the documents which the contract data states it is in.

Unlike its frequent references to works information, the text of the NEC contains very few references to site information. They are:

- clause 54.1 – information in the activity schedule is not site information
- clause 55.1 – information in the bill of quantities is not site information
- clause 60.2 – the contractor is assumed to have taken into account site information in judging physical conditions
- clause 60.3 – if there is inconsistency in the site information the contractor is assumed to have taken account of the physical conditions most favourable to doing the work.

Only the last two of these are significant. Their significance derives from compensation event clause 60.1(12), which comes into effect when the contractor encounters physical conditions within the site which an experienced contractor would have adjudged to have had such a small chance of occurring that it would have been unreasonable for him to have allowed for them.

## 4.7 Site information

There is no express requirement in the NEC for the employer to supply all or any of the information he has on the site. Nor is there any express requirement on the contractor to satisfy himself as to conditions on the site.

### Importance of site information

How important is the site information in the operation of the NEC? It has been suggested in some quarters that the NEC has a different policy from other contracts in that the contractor is entitled to rely on the site information provided by the employer in order to price the work. But even if that is the policy, does it work?

To test this assume that the employer fails to provide any site information (even though he may have some). The contractor encounters unforeseen physical conditions and claims under compensation event 60.1(12). The judgment now to be made under clause 60.2 is solely on:

- information obtainable from a visual inspection of the site, *and*
- other information which an experienced contractor could reasonably be expected to have or to obtain.

This is a long way from saying that the contractor is entitled to rely on the site information. And it is certainly not as favourable to the contractor as the ICE 6th edition conditions of contract which expressly require the employer to make available all information which he has on the site.

### Identification of site information

The statement in clause 11.2(6) that site information is in the documents which the contract data states it is in, may not be of much effect. Any site information provided to the contractor (whether mentioned in the contract data or not) would have to be taken into account under clause 60.2.

### Inconsistency within the site information

Clause 60.3 provides that if there is any inconsistency within the site information the contractor is assumed to have taken into account the physical conditions more favourable to doing the work. See the comment in Chapter 7, section 2, on the dangers of applying this provision to contractor's design.

**Changes in site information**

The NEC does not allow for any changes in the defined site information. So for contractual purposes the site information is fixed at the time of tender.

In practice, of course, there may be changes on the site as the works progress. The employer, for example, may have other contractors installing pipes and the like or additional information on the site may come to light. The employer may consider it good practice to convey this new information to the contractor and, indeed, may have a duty to do so on safety grounds. The question then is – how does this new information fit into the contractual framework of the NEC?

In some cases it may be necessary for the project manager to give instructions changing the works information to accommodate this new site information. But without an instruction the contractor may have to rely on compensation event 60.1(12). But this is arguably not applicable to supervening events.

## 4.8 Contract data

Part one of the contract data is provided by the employer, initially to tenderers. It serves the purpose of the appendix to the form of tender in detailing specifics relating to the contract.

Part two of the contract data is provided by the contractor, and is returned with his tender. It details specifics relating to his company and his tender.

The model forms incorporated within the NEC cover all the necessary entries, but if the forms are simply used as printed it will be found that more space is required for some of the entries.

## 4.9 Schedules of cost components

The NEC contains two schedules of cost components:

- the schedule of cost components
- the shorter schedule of cost components.

In order to incorporate these schedules into the contract it is probably best to refer to them expressly because their inclusion may not automatically follow from inclusion of the contract data or the NEC conditions.

## 4.9 Schedules of cost components

**Purpose of the schedules**

In simple terms the schedules provide a division between reimbursable items of cost and the items deemed to be included in the contractor's fee when payments to the contractor are to be calculated on a cost basis.

The application of cost schedules to Options C, D and E is obvious enough. These options are of a cost reimbursable nature. In Options A and B, which are essentially firm price contracts, the only use of the cost schedules is in the assessment of compensation events, most of which have to be assessed with regard to forecast actual cost.

Both of the schedules commence with a statement that they do not form part of the conditions of contract when Option F (the management contract) is used. This is understandable when the contractor subcontracts all the work – which seems to be the general intention. But it is not clear why the schedules should not apply if the contractor does some of the work himself, as seems to be permitted by clause 20.2.

**Use of the shorter schedule**

Use of the shorter schedule is confined to the assessment of compensation events, and only then when there is either agreement between the project manager and the contractor on its use or when the project manager makes his own assessments (clause 63.11).

**Differences between the schedules**

The main schedule of cost components is a fully detailed schedule under which all costs have to be evidenced except the contractor's fee. The fee itself is tendered and comprises mainly head office overheads, insurances, taxes, and the like. Because of the amount of detail involved, the process of calculating amounts due by the main schedule requires considerable time and effort.

The shorter schedule provides a simpler method of assessment with:

- fewer cost components
- equipment costs based on rates rather than depreciation and maintenance
- miscellaneous charges and consumables covered by a tendered percentage.

**Data provided by the contractor**

At the tender stage the contractor is required to price only a limited amount of information for the cost schedules. The requirements are detailed in part two of the contract data. Contrary to the thoughts of some early users of the NEC the contractor is certainly not required to price at tender stage all the various components of cost.

The key item which has to be provided is the 'fee percentage'. The fee has to cover everything which is not in the cost components.

Other items which have to be priced at tender stage are:

- percentages for depreciation and maintenance of equipment. These are used in calculations of cost for equipment covered by the contractor. Note that the stated percentages do not apply to the shorter schedule which values equipment costs on a rates basis
- a percentage for working area overheads. This is a percentage applied to the cost of people which is to cover the list of miscellaneous on-site overheads stated in the main schedule of cost components. Again this percentage does not apply to the shorter schedule
- hourly rates and a percentage for overheads for costs of manufacture and fabrication outside the working areas
- hourly rates and a percentage for overheads for design costs outside the working areas.

For the shorter schedule of cost components the items to be priced in the tender are:

- a percentage for 'people' overheads. This is applied to wages and salaries etc. to cover the list of miscellaneous costs in paragraph 4 of the shorter schedule
- a percentage for adjustment of the rates for listed equipment
- rates for non-listed equipment.

The various rates and percentages entered by the contractor in the tender do not carry through into the tender total and employers are left to decide for themselves what procedure to adopt to ensure that they are competitive. The common approach is to apply the rates and percentages to notional figures for comparison of tenders. But policies vary on whether or not to inform the tenderers of the amounts of such figures.

**Costs of subcontractors**

The manner in which the costs of subcontractors are dealt with under the schedules of cost components may in some instances be a matter of concern to contractors.

The problem is that for Options A and B the term 'contractor' includes for 'subcontractors'. This is not stated openly in the schedules but it follows from the opening statement in both schedules that when Options C,

D or E are used the 'contractor' means the 'contractor' and not his 'subcontractors'. In other words only for Options C, D and E are subcontractor costs recoverable on an invoice basis under the schedules. They are, of course, also recoverable for Option F – but the schedules are stated not to apply to that option.

For Options A and B, however, any subcontractor costs are priced out using the cost component method and the only percentage uplift allowed is the contractor's fee. In short, there is no uplift for the subcontractor's overheads and they are deemed to be included in the contractor's fee.

What this scheme seems to overlook is that subcontractor's overheads, particularly when the subcontractors are specialist suppliers, will usually be much higher than the main contractor's overheads. A fee of 5% might be reasonable for a main contractor but 15% to 25% might be justifiable for subcontractors.

It is not obvious how this difficulty can be overcome – unless contractors quote very high fees. But, having regard to the fact that the avowed purpose of the compensation event procedure of the NEC is to 'compensate' the contractor, and having regard to the obligation on the employer and project manager under clause 10.1 of the contract to act in a spirit of mutual trust and co-operation, it would seem to be incumbent on the project manager to find a way out of the difficulty.

**Contractor's own equipment**

Another aspect of the schedules of cost components which contractors need to watch carefully is the way in which the contractor's own equipment is costed. Under the main schedule, hire rates do not apply to equipment which is owned:

- by the contractor
- by the contractor's parent company
- by any part of a group with the same parent company.

Equipment which is so owned is costed on a depreciation and maintenance basis.

Under the shorter schedule, however, no distinction is made between hired plant and owned plant and both are costed on a rates basis.

The outcome is likely to be that in many cases (if not most) the contractor who owns equipment will be better off with an assessment made under the shorter schedule. This will particularly apply if some element of the contractor's claim is delay related. There does not appear to be much scope (if any) under the main schedule for recognising loss of opportunity to use owned plant elsewhere as a legitimate cost.

## 4.10 Ambiguities and inconsistencies in the contract documents

Clause 17.1 of the NEC deals with ambiguities and inconsistencies in the contract documents. It requires the project manager or the contractor to notify the other as soon as either becomes aware of any ambiguity or inconsistency in or between the documents forming the contract. The project manager is then required to give an instruction resolving the matter.

At first sight clause 17.1 looks very much like the usual 'ambiguities' clause found in other standard forms. But on closer examination it can be seen to be much wider in its scope than is normal. Most 'ambiguities' clauses deal only with matters relating to the construction of the works; they do not normally extend to interpretation (or alteration) of the contract itself.

Clause 17.1, however, is very widely drafted. Taken literally it would lead to absurd results, such as all the questions in this book being landed on the project manager's lap for resolution. But even given a narrow common sense application the scope of the clause is still worrying. It appears to go well beyond authorising instructions to change the works information.

And even for changes to works information there has to be some concern since the clause appears to put the responsibility of resolving ambiguities and inconsistencies in the contractor's design on the project manager. This cannot be what is intended and project managers would be most unwise to take on the burden of sorting out problems in the contractor's design.

## 4.11 Schedule of clauses referring to the works information

### 1 Common core clauses

| | |
|---|---|
| Clause 11.2(5) | – definition of works information |
| Clause 11.2(11) | – equipment is defined by reference to what the works information does not require to be included in the works |
| Clause 11.2(13) | – completion is when the contractor has done all the work the works information states is to be done by the completion date |
| Clause 11.2(15) | – a part of the works not in accordance with the works information is a defect |
| Clause 14.3 | – the works information can be changed by instruction of the project manager |

## 4.11 Schedule of clauses referring to the works information

| | |
|---|---|
| Clause 18.1 | – the contractor is to act in accordance with health and safety requirements stated in the works information |
| Clause 19.1 | – if the contractor is required by the works information to do anything illegal or impossible the contractor gives notice and the project manager instructs appropriately |
| Clause 20.1 | – the contractor is to provide the works in accordance with the works information |
| Clause 21.1 | – the contractor is to design such parts of the works as stated in the works information |
| Clause 21.2 | – the contractor is to submit for acceptance such particulars of his design as the works information requires |
| | – a reason for the project manager not accepting the design is that it does not comply with the works information |
| Clause 22.1 | – the employer's entitlement to use the contractor's design may be restricted or expanded in the works information |
| Clause 23.1 | – a reason for the project manager not accepting the design of an item of equipment is that it will not allow the contractor to provide the works in accordance with the works information |
| Clause 25.1 | – the contractor is to share the working areas with others as stated in the works information |
| Clause 31.2 | – the contractor is to show on each programme the work of the employer and others as stated in the works information |
| | – the contractor is to show on each programme any other information which the works information requires |
| Clause 31.3 | – a reason for the project manager not accepting a programme is that it does not comply with the works information |
| Clause 33.2 | – the employer and the contractor are to provide such facilities and services as are stated in the works information |
| Clause 35.3 | – the employer is deemed to take over any part of the works he uses before completion unless the use is for a reason stated in the works information |
| Clause 40.1 | – clause 40 applies only to tests and inspections required by the works information (and the applicable law) |

## 4.11 Schedule of clauses referring to the works information

| | |
|---|---|
| Clause 40.2 | – the contractor and the employer are to provide materials, facilities and samples for tests and inspections as stated in the works information |
| Clause 42.1 | – searching may include doing tests and inspections which the works information does not require |
| Clause 44.1 | – the works information may be changed so that a defect does not have to be corrected |
| Clause 44.2 | – the project manager gives an instruction changing the works information if he accepts that a defect need not be corrected |
| Clause 60.1(1) | – an instruction changing the works information is a compensation event unless it is:<br>○ a change to accept a defect<br>○ a change to the contractor's design at his request or to comply with other works information |
| Clause 60.1(5) | – failure by the employer or others to work within the times stated in the works information is a compensation event |
| Clause 60(1)(16) | – failure by the employer to provide materials, facilities and samples for tests as stated in the works information is a compensation event |
| Clause 63.2 | – the prices are reduced if the effect of a compensation event is to reduce the actual cost and the event is a change to the works information |
| Clause 63.7 | – the assessment of a compensation event which is an instruction to change the works information in order to resolve an ambiguity or inconsistency, is made having regard to which party provided the works information |
| Clause 71.1 | – payment for plant and materials outside the working areas is dependent upon the contractor preparing them for marking as the works information requires |
| Clause 73.2 | – the contractor has title to materials from excavation and demolition only as stated in the works information. |

### 2 Clauses in Options A & C

| | |
|---|---|
| Clause 54.1 | – information in the activity schedule is not works information. |

*4.11 Schedule of clauses referring to the works information* 65

## 3 Clauses in Options B & D

Clause 55.1      – information in the bill of quantities is not works information.

## 4 Clauses in Options C, D, E & F

Clause 52.2      – the contractor is to keep accounts and records as stated in the works information.

## 5 Clauses in Options C, D & E

Clause 11.2(30)      – cost incurred because the contractor did not follow an acceptance or procurement procedure stated in the works information, is a disallowed cost
– cost of correcting a defect caused by the contractor not complying with the works information, is a disallowed cost.

## 6 Clauses in Options C & D

Clause 53.5      – the prices are not reduced if the project manager accepts a proposal by the contractor to change the works information provided by the employer.

## 7 Clauses in Option F

Clause 11.2(29)      – cost incurred because the contractor did not follow an acceptance or procurement procedure stated in the works information, is a disallowed cost

Clause 20.2      – the contractor is to subcontract design, construction, installation and such other work as stated in the works information
– the contractor may either subcontract work not so stated in the works information or do it himself.

## 8 Secondary option clauses

| | |
|---|---|
| Performance bond G1.1 | – the contractor gives a performance bond in the form set out in the works information |
| Parent company H1.1 | – the contractor gives a parent company guarantee in the form set out in the works information (if applicable) |
| Advanced payment J1.2 | – the bond for advanced payment is in the form set out in the works information |
| The contractor's design M1.1 | – the contractor is not liable for defects in his design if he proves he has used reasonable skill and care to ensure it complied with the works information. |

# Chapter 5

# The key players

## 5.1 Introduction

The NEC states in various degrees of detail the obligations, duties and powers of the following participants:

- employer
- contractor
- project manager
- supervisor
- adjudicator.

Additionally the NEC has something to say on subcontractors, suppliers and 'others'. Nowhere, however, does the NEC mention the designer as such. His function falls on the employer or the contractor according to which party is responsible for design.

### The parties

Clause 11.2(1) states simply that the parties to the contract are the employer and the contractor. Both are terms to be identified by name in the contract data. Only the term 'the Parties' is to be regarded as a defined term.

### The project manager, supervisor and adjudicator

Again these are identified terms and not defined terms. All are to be named in the contract data.

The project manager and the supervisor can be replaced by the employer but only after the employer has notified the contractor of the name of any replacement (clause 14.4). There is apparently no need for consultation as with some model forms.

### Subcontractors

Clause 11.2(9) makes 'subcontractor' a defined term. Subcontractors are not required to be identified in the contract data but the contractor must submit their names for acceptance by the project manager before their appointment (clause 26.2).

The definition of subcontractor in clause 11.2(9) is not the easiest to follow. It states that a subcontractor is:

- a person or corporate body
- who has a contract with the contractor
- to provide part of the works, *or*
- to supply plant and materials which he has wholly or partly designed specifically for the works.

It is the last part of the clause which requires some examination. Why should a person or body who has supplied plant or materials which he has wholly or partly designed for the works be a subcontractor and not a supplier? The answer, perhaps, is that the NEC exercises no control over suppliers. They do not come within the scope of clause 26 on subcontracting and they are not treated in the same way as subcontractors in the schedules of cost components.

Presumably, therefore, the intention of the NEC is that suppliers who design specifically for the works should come within its control mechanisms and that is why they are defined as subcontractors. There is, however, a potential difficulty with this. Such firms may have no wish to trade as subcontractors rather than suppliers. And it may be restrictive on the contractor's choice of specialist firms and also on the designers to insist that they do.

**Obligations and responsibilities**

It is not intended in this chapter to provide detailed comment on the involvement, obligations and responsibilities under the NEC of all its participants. Only the employer, the project manager, the supervisor and others are covered here.

For comment on the contractor see Chapter 7; for comment on subcontractors see Chapter 16; for comment on designers see Chapter 17; and for comment on the adjudicator see Chapters 14 and 18.

## 5.2 Others

In the execution of most contracts the parties are likely to have dealings with many persons and organisations who are, in a legal sense, remote from the contract itself. The circumstances of such dealings are potentially so varied that standard forms of contract do not usually attempt to state the contractual effects of the dealings. Most contracts rely on the application of the common law rule of prevention that neither party must prevent the other from undertaking the fulfilment of its obligations. And to the extent that the parties are responsible for others, the rule extends so that the parties are responsible for prevention caused by others.

The NEC ventures further than most standard forms in setting out the obligations and liabilities of the parties in respect of remote persons and organisations. Firstly, it defines them as 'others' and then in six of its clauses it goes on to state their impact on the contract. The intention appears to be to stay fairly close to the legal rule of prevention, but as will be seen from the comment below the generality of the definition could lead to some unintended effects.

**Definition**

Clause 11.2(2) defines the term 'Others' by exclusion. It states that 'Others' are people or organisations who are not the employer, project manager, supervisor, adjudicator, contractor, or any employee, subcontractor or supplier of the contractor.

An interesting aspect of this clause is its specific reference to 'any employee' of the contractor. Under the rules for the construction of contracts it may follow that employees of the other categories mentioned in the definition and who are not so specifically excluded are to be taken as 'Others'.

**Express references to 'Others'**

Except where there are express references to 'others' in the NEC, common law rules will apply to their impact on the contract. The express references are as follows:

- clause 25.1    – the contractor is to co-operate with others in obtaining and providing information they need in connection with the works
                 – the contractor is required to share the working areas with others as stated in the works information
- clause 27.1    – the contractor is responsible for obtaining approval of his design from others where necessary
- clause 31.2    – the contractor is required to show on his programme the work of others as stated in the works information, and dates relating thereto
- clause 60.1(5) – failure by others to work within the times or conditions stated in the works information is a compensation event

- clause 80.1 – loss or damage to plant or materials supplied by others on the employer's behalf is an employer's risk until the contractor receives them
- clause 95.3 – a reason for termination of the contract by the employer is that the contractor has substantially hindered others.

**Application**

Common sense is obviously required in the application of these clauses. Taken literally their scope is very wide. For example, the contractor would apparently have to co-operate under clause 25.1 with a journalist sent to write an article on the works, and his failure to do so would arguably be grounds for termination under clause 95.3. But this can be avoided by implying a qualification of relevance into the definition of others or by implying a test of relevancy to the application of the clauses.

Note, however, that whereas the impact of others is specifically defined by reference to the works information, the contractual provisions are effectively silent on the wider impact of others.

## 5.3 Actions

The opening clause of the NEC, clause 10.1, has been much debated. Within the style of the NEC the clause is unusual in that it uses the word 'shall'. Within the realm of contracts generally the clause is unusual in that it expressly requires the parties to act in a spirit of mutual trust and co-operation.

**Clause 10.1**

Clause 10.1 states that:

- the employer, contractor, project manager and supervisor 'shall' act as stated in the contract, *and*
- in a spirit of mutual trust and co-operation.

The clause also states that:

- the adjudicator shall act as stated in the contract, *and*
- in a spirit of independence.

## Obligation to act

A unique feature of the NEC is that generally it avoids the convention of using the word 'shall' to indicate obligations. Its style relies on present tense verbs such as 'notifies', 'provides', 'submits' to convey the intention that an obligation exists. In the very first line, however, of the first clause in the NEC the phrase 'shall act' appears. The intention appears to be that this 'shall' acts for all following clauses such that a clause 'the Contractor notifies' is to be read as meaning 'the Contractor shall notify'. However, as some lawyers have pointed out, the use of the word 'shall' in one clause and its omission from others could have the effect of casting doubt on how some clauses of the contract are to be interpreted.

## Failure to act

Failure by the employer or the contractor to act as stated in the contract is a breach of contract for which there may either be remedies in the contract or at common law.

In the NEC the majority of listed contractual remedies are called 'compensation events' and with only minor exceptions these are available only to the contractor. There are, however, various other express remedies distributed throughout the NEC and many of these are available to the employer, e.g. damages for delay, termination, recovery of certain costs, withholding a proportion of payments due. Compared with other standard forms the NEC has a wide range of such remedies.

Where there are no contractual remedies and the parties rely on their common law rights, a common problem, particularly for the employer, is that the proof of loss necessary to recover damages is often difficult, if not impossible, in respect of some breaches of contract – especially procedural ones.

The question of whether failure by the project manager or supervisor to act as stated in the contract is a breach of contract, is quite complex. Neither of them is a party to the contract and the question applying generally to contracts is to what extent the employer does warrant the competency and/or fairness of any contract administrator or supervisor appointed by him.

In the NEC that question is largely avoided by a precise set of rules and remedies. But it is not avoided altogether and for further comment see section 5.6 below, on the project manager.

As to the obligation in clause 10.1 that the adjudicator shall act as stated in the contract, this is most unusual and hard to follow in its purpose or effect. It can hardly be intended that the employer is responsible for any failure by the adjudicator to act as stated – although that might be one interpretation of the clause. If the adjudicator does fail to act as stated in the contract that would, in the normal course of things, be a breach of any contracts between the adjudicator and the parties and nothing to do with the main contract itself.

**Mutual trust and co-operation**

The NEC has taken a brave step forward, so far as English law is concerned, in expressly including (in clause 10.1) an obligation for the parties to act in 'a spirit of mutual trust and co-operation'.

This is very much a step into the unknown, for as Mr Justice Vinelott said in the case of *Merton* v. *Leach* (1985), 'the courts have not gone beyond the implication of a duty to co-operate wherever it is reasonably necessary to enable the other party to perform his obligations under a contract. The requirement of "good faith" in systems derived from Roman law has not been imported into English law'. And twenty years earlier Lord Devlin in his lecture 'Morals and the Law of Contract' had made the telling point that, 'If a man minded only about keeping faith, the spirit of the contract would be more important than the letter. But in the service of commerce the letter is in many ways the more significant'.

It is, of course, true that many jurisdictions outside England do impose duties of good faith and fair dealing into contracts. But it is questionable whether, when it comes to enforcement, these duties amount to much more than a duty not to act in bad faith. And that, perhaps, is little different from the principle well recognised in English law that prevention is a breach of contract.

The difficulty with such admirable concepts as good faith, fair dealing, mutual trust and co-operation is determining what function they are intended to serve. Are they really intended to be legally enforceable, with all the attendant difficulties of proof of breach and proof of loss? Or are they no more than expressions of good intent or warranties to act reasonably? It is not easy to determine from the wording of clause 10.1 which of the above the NEC intends.

If the clause said (which it does not), 'The Employer, the Contractor, the Project Manager and the Supervisor shall act as stated in this contract in a spirit of mutual trust and co-operation', then its purpose would be reasonably clear. It would qualify the rights of the parties, the project manager and the supervisor to apply the provisions of the NEC without regard to a test not far dissimilar to one of reasonableness; although for important matters such as termination, where such a qualification would have its greatest effect, it is worth noting that the law may already imply a test of reasonableness. See the Australian case of *Renard Constructions* v. *Minister of Public Works* (1992).

But what clause 10.1 actually says is 'The Employer, the Contractor, the Project Manager and the Supervisor shall act as stated in this contract and in a spirit of mutual trust and co-operation'. There are, therefore, two distinct obligations in the clause:

- to act as stated in the contract 'and'
- to act in a spirit of mutual trust and co-operation.

This suggests an intention that the provisions of the contract are to be strictly applied and that there is a further obligation outside the stated provisions which is of separate application. But the problem with this approach is that, whilst it might conceivably apply to the employer and to the contractor, it is difficult to see how it could ever apply to the project manager and to the supervisor. They have only one role and that is to apply the provisions of the contract.

The debate on the proper legal construction of clause 10.1 will probably run until it is settled in the courts. But in the meantime contractors rather than employers are likely to be the main beneficiaries of any confusion. They will have more opportunities to capitalise on uncertainty as to whether it applies to individual provisions. And they will be better placed to prove damages in a claim for breach.

For comment on the application of clause 10.1 to the adjudicator see Chapter 18, section 4.

## 5.4 The employer

The role of the employer in the NEC is strictly that of a legal party. The employer is not intended to have any direct involvement in the running of the contract, except through the project manager or in the rare event of termination. Two clauses which emphasise this are:

- clause 14.3 – only the project manager may give an instruction which changes the works information
- clause 29.1 – the contractor is required to obey instructions given only by the project manager or the supervisor.

That is not to say, however, that the employer under an NEC contract has less influence on the running of the contract than with other standard forms. If anything the position is the reverse. The project manager under the NEC is clearly the employer's agent for many of his functions and the employer is entitled to exercise more influence on him than would be permissible with an independent contract administrator.

### Contractual obligations

The primary obligations of the employer in any construction related contract are to allow possession of the site to the contractor and to pay the contractor. The NEC contains these, of course, and it also contains a variety of secondary obligations. See the schedule in section 5.5 below.

Failure by the employer to comply with any of his obligations is a breach of contract which either entitles the contractor to a compensation event under clause 60 or damages at common law. For further comment on this see Chapter 11.

A point which has been noted by a number of commentators on the NEC is the absence of any express obligation on the employer to provide (through the project manager) information, additional to that in the works information, necessary for the contractor to complete the works. The question that poses is whether the employer is bound by an implied term to provide the missing information, or whether the contractor has assumed the obligation to fill any gaps himself. The answer will vary according to the facts of each case.

## 5.5 Express obligations of the employer

**Core clauses**

| | |
|---|---|
| 10.1 | – to act as stated in the contract and in a spirit of mutual trust and co-operation |
| 14.4 | – to give notice to the contractor before replacing the project manager or the supervisor |
| 33.1 | – to give possession of each part of the site before the later of the possession date and the date for possession shown on the accepted programme |
| 33.2 | – to give the contractor access to the site |
| | – to provide facilities and services as stated in the works information |
| 35.2 | – to take over the works not more than two weeks after completion |
| 35.3 | – to take over any part of the works put into use (subject to exceptions) |
| 40.2 | – to provide materials, facilities and samples for tests and inspections as stated in the works information |
| 43.3 | – to give access to the contractor after take over if needed for the correction of a defect |
| 51.1 | – to pay amounts due to the contractor |
| 51.2 | – to pay within three weeks of the assessment date or to pay interest on late payment |
| 83.1 | – to indemnify the contractor against claims etc. from employer's risks |
| 84.1 | – to provide insurances as stated in the contract data |
| 85.3 | – to comply with the terms and conditions of insurance policies |

## 5.5 Express obligations of the employer

94.1 – to notify the project manager giving reasons before terminating.

**Secondary option clauses**

*Option J1 – advanced payment*

J1.1 – to make the advanced payment of the amount stated in the contract data
J1.2 – to make the advanced payment within four weeks of the contract date or receipt of the advanced payment bond.

*Option R – delay damages*

R1.2 – to repay any overpayment of delay damages with interest.

*Option V – trust fund*

V2.1 – to establish the trust fund within one week of the contract date
V3.1(4) – to restore the trust fund to its initial value
V3.1(6) – to pay the trustees their fees and expenses.

## 5.6 The project manager

The role of the project manager in an NEC contract is an involved and demanding one. The contractor and the project manager are intended to work together to see the contract through to completion. And, for the employer, the successful outcome of the contract will depend on the competence of the project manager.

The full extent of the project manager's duties can be seen from the schedule in section 5.7 below. The list is long by any standards.

**Appointment**

Some employers will have within their organisations persons of sufficient ability and experience to take on the role of project manager and there is nothing in the NEC to prevent an in-house appointment. Other employers, through necessity or choice, will appoint an external firm or person as the project manager.

All that the NEC requires in contractual terms is that the employer should state the name of the project manager in part one of the contract data and that the employer should not replace the project manager before giving notice to the contractor of the name of the replacement (clause 14.4).

### Firm or person

The NEC places no restrictions on who or what the project manager should be. Unlike the engineer in the ICE 6th edition conditions of contract, he is not required to be a named individual. Nor is he required to be a chartered engineer.

With external appointments the project manager will more often than not be named as a firm, if only to secure the cover of the firm's professional indemnity insurance. But for operational purposes there is much to be said for the identification of a particular person as the project manager.

### Delegation

Clause 14.2 permits the project manager to delegate any of his actions. The only formality is that the contractor should be notified.

There are no restrictions in the NEC on delegation either in regard to which duties and powers can or cannot be delegated, or in regard to how many delegates there can be. This could lead to a dangerous division of responsibility and prudent employers will no doubt impose their own restrictions on delegation to avoid being caught by its consequences.

### Impartiality and fairness

There is no express requirement in the NEC for the project manager to be impartial and much that has been written about the NEC assumes that it dispenses entirely with the need for impartiality.

The proposition is advanced that the project manager is required to concern himself solely with the interests of the employer, for whom he acts as agent, and that it is the task of the adjudicator to step in and resolve any differences with the contractor. Some parts of the NEC support this – most notably, perhaps, the fact that under the disputes resolution procedures in Section 9, the employer has no right to challenge any of the actions of the project manager or the supervisor. This does suggest that the actions of the project manager are taken on behalf of the employer.

But there are parts of the NEC which do not work if the project manager simply puts the interests of the employer before impartiality. For example, the termination provisions in Section 9 require the project manager to issue a termination certificate (or to reject with reasons) on the application of either party. It cannot be the case that in considering any application the project manager should put the interests of the employer, who may be the guilty party, as paramount.

And even on a more routine scale of decision making such as issuing certificates, valuing compensation events and the like, it is difficult to see why the project manager, when exercising his discretion as a certifier or valuer, should be excused the ordinarily legal requirement to act with fairness. Failure to do so would surely put the employer in breach of the well established implied term that his appointed certifier or valuer should act fairly; and it might expose the project manager to an action in tort from the contractor.

## 5.7 Express duties of the project manager

**Common core clauses**

10.1 – to act as stated in the contract and in a spirit of mutual trust and co-operation
13.1 – to communicate in a form which can be read, copied and recorded
13.3 – to reply to a communication within the period for reply
13.4 – to reply to a communication submitted or resubmitted for acceptance
  – to state reasons for non-acceptance
13.5 – to notify any agreed extension to the period for reply
13.6 – to issue certificates to the employer and to the contractor
13.7 – to communicate notifications separately from other communications
16.1 – to give early warning of matters with delay, cost or performance implications
16.3 – to co-operate at early warning meetings
16.4 – to record proposals considered and decisions taken at early warning meetings
  – to give a copy of the record to the contractor
17.1 – to give notice of ambiguities or inconsistencies in the documents
  – to give instructions resolving ambiguities or inconsistencies
19.1 – to give instructions changing the works information in the event of illegality or impossibility in the works information
21.2 – to accept particulars of the contractor's design or to give reasons for non-acceptance
23.1 – to accept particulars of the design of equipment or to give reasons for non-acceptance
24.1 – to accept replacement persons proposed by the contractor or to give reasons for non-acceptance

26.2 – to accept proposed subcontractors or to give reasons for non-acceptance
26.3 – to accept proposed subcontract conditions or to give reasons for non-acceptance
30.2 – to decide the date of completion
 – to certify completion within one week of completion
31.3 – to accept the contractor's programme within two weeks of submission or to give reasons for non-acceptance
32.2 – to accept a revised programme or to give reasons for non-acceptance
33.2 – to assess any cost incurred by the employer as a result of the contractor not providing facilities and services as stated in the works information
35.4 – to certify within one week the date when the employer takes over any part of the works
40.6 – to assess the cost incurred by the employer in repeating a test or inspection after a defect is found
43.3 – to arrange for the employer to give access and use to the contractor of any part of the works needed for the correction of defects after taking over
 – to extend the period for correcting defects if suitable access and use is not arranged within the defect correction period
44.2 – to change the works information, the prices and the completion date if a quotation for not correcting defects is accepted
45.1 – to assess the cost of having defects corrected by others if the contractor fails to correct notified defects within the defect correction period
50.1 – to assess the amount due for payment at each assessment date
 – to decide the first assessment date to suit the procedures of the parties
50.4 – to consider any application from the contractor when assessing amounts due for payment
 – to give the contractor details of how amounts due have been assessed
51.1 – to certify payment within one week of each assessment date
61.1 – to notify the contractor of compensation events which arise from the giving of instructions or changing of earlier decisions
 – to instruct the contractor to submit quotations
61.4 – to decide within one week of notification (or such longer period as the contractor agrees) whether the prices and the completion date should be changed when the contractor notifies a compensation event
 – to instruct the contractor to submit quotations for changed prices and the completion date
61.5 – to decide whether the contractor did not give any early warning

## 5.7 Express duties of the project manager

of a compensation event which should have been given and to notify the contractor of his decision
- 61.6 – to state assumptions for the assessment of compensation events in the event that the effects are too uncertain to be forecast reasonably
  – to correct any assumptions later found to have been wrong
- 62.3 – to reply within two weeks to quotations for compensation events submitted by the contractor
- 62.4 – to give reasons to the contractor when instructing the submission of a revised quotation
- 62.5 – to extend the time allowed for the submission of quotations and replies if the contractor agrees
  – to notify the contractor of any agreed extensions for the submission of quotations or replies
- 64.1 – to assess a compensation event:
  ○ if the contractor has not submitted a quotation within the time allowed
  ○ if the project manager decides the contractor has not assessed the compensation event correctly
  ○ if the contractor has not submitted a required programme
  ○ if the project manager has not accepted the contractor's latest programme
- 64.2 – to assess a compensation event using his own assessment of the programme:
  ○ if there is no accepted programme
  ○ if the contractor has not submitted a revised programme for acceptance as required
- 64.3 – to notify the contractor of any assessments made of a compensation event within the period allowed to the contractor for his quotation
- 65.1 – to implement compensation events by notifying the contractor of accepted quotations or his own assessments
- 73.1 – to instruct the contractor how to deal with objects of value, historical or other interest
- 85.1 – to accept policies and certificates of insurance submitted by the contractor or to give reasons for non-acceptance
- 87.1 – to submit to the contractor policies and certificates for insurances to be provided by the employer
- 90.2 – to proceed with matters in dispute which are referred to adjudication as though they were not disputed until there is a settlement
- 94.1 – to issue a termination certificate when either party gives notice of termination for reasons complying with the contract
- 94.4 – to certify final payments within 13 weeks of termination.

**Main option clauses**

*Option A*

36.3 – to change the completion date and prices when a quotation for acceleration is accepted
– to accept the revised programme
54.2 – to accept a revision to the activity schedule or to give reasons for non-acceptance
65.4 – to include changes to the prices and to the completion date when notifying implementation of a compensation event.

*Option B*

36.3 – to change the completion date and prices when a quotation for acceleration is accepted
60.6 – to correct mistakes in the bill of quantities
65.4 – to include changes to the prices and to the completion date when notifying implementation of a compensation event.

*Option C*

26.4 – to accept proposed contract data for subcontracts or to give reasons for non-acceptance
36.3 – to change the completion date and prices when a quotation for acceleration is accepted
36.5 – to accept a subcontractor's proposal to accelerate or to give reasons for non-acceptance
53.1 – to assess the contractor's share for interim payments
53.3 – to assess the contractor's share at completion
53.4 – to assess the contractor's share in the final amount due
54.2 – to accept a revision to the activity schedule or to give reasons for non-acceptance
65.4 – to include changes to the prices and to the completion date when notifying implementation of a compensation event
97.4 – to assess the contractor's share after certifying termination.

*Option D*

26.4 – to accept proposed contract data for subcontracts or to give reasons for non-acceptance
36.3 – to change the completion date and prices when a quotation for acceleration is accepted

36.5 – to accept a subcontractor's proposal to accelerate or to give reasons for non-acceptance
53.1 – to assess the contractor's share for interim payments
53.3 – to assess the contractor's share at completion
53.4 – to assess the contractor's share in the final amount due
60.6 – to correct mistakes in the bill of quantities
65.4 – to include changes to the prices and to the completion date when notifying implementation of a compensation event
97.4 – to assess the contractor's share after certifying termination.

## Option E

26.4 – to accept proposed contract data for subcontracts or to give reasons for non-acceptance
36.4 – to change the completion date when a quotation for acceleration is accepted
36.5 – to accept a subcontractor's proposal to accelerate or to give reasons for non-acceptance
65.3 – to include changes to the forecast amount of the prices and the completion date when implementing a compensation event.

## Option F

26.4 – to accept proposed contract data for subcontracts or to give reasons for non-acceptance
36.4 – to change the completion date when a quotation for acceleration is accepted
36.5 – to accept a subcontractor's proposal to accelerate or to give reasons for non-acceptance
65.3 – to include changes to the forecast amount of the prices and the completion date when implementing a compensation event.

**Secondary option clauses**

## Option G – performance bond

G1.1 – to accept the contractor's performance bond or to give reasons for non-acceptance.

## Option J – advanced payment to the contractor

J1.2 – to accept an advanced payment bond or to give reasons for non-acceptance.

## 5.8 The supervisor

One of the most unusual features of the NEC is the way it separates the functions of contract administration and supervision. It does this not by delegation of powers but by specifying different roles for the project manager and supervisor and by apparently empowering them with independence from each other.

The express duties of the supervisor are listed in section 5.9 which follows and from this it can be seen that the supervisor is mainly concerned with the quality of work and defects.

### Appointment

The NEC requires the name of the supervisor to be entered by the employer in part one of the contract data. And like the project manager, the supervisor can be either a firm or a person. There is no express prohibition in the NEC on the supervisor and the project manager being the same firm or person and there is no serious procedural difficulty if that is the case.

Practical difficulties are, perhaps, more likely to arise when the project manager and the supervisor come from different organisations and each takes an uncompromising position on his duties under the contract. An over zealous supervisor could, for example, upset the project manager's plans to put early completion as top priority.

And note in particular that although the supervisor issues the defects certificate (clause 43.2), decisions on accepting defects are taken by the project manager (clause 44.1). Note also that the supervisor has no express role to play in the termination procedures set out in Section 9 of the NEC.

### Delegation

Under clause 14.2 the supervisor can delegate his duties without restriction.

## 5.9 Express duties of the supervisor

### Core clauses

- 10.1 – to act as stated in the contract and in a spirit of mutual trust and co-operation
- 13.1 – to communicate in a form which can be read, copied and recorded

13.3 – to reply to a communication within the period for reply
13.6 – to issue certificates to the project manager and the contractor
40.3 – to notify the contractor of his tests and inspections before they start and afterwards of the results
40.5 – to do tests and inspections without causing unnecessary delay
42.1 – to give reasons for searches which are instructed
42.2 – to notify the contractor of defects found
43.2 – to issue the defects certificate
71.1 – to mark equipment, plant and materials outside the working areas for payment purposes.

## 5.10 Communications

The NEC states in much greater detail than any other standard form of contract how communications between the contractor and the project manager or the supervisor are to be conducted. The requirements themselves are set out in clause 13 but equally important to the contractor is clause 60.1(6) – the compensation event for failure by the project manager or the supervisor to reply to a communication in time.

Note that communication is not a defined term of the NEC and that its meaning has to be gathered from the contract.

**Clause 13.1 – communications**

The main purpose of clause 13.1 appears to be to list those documents/exchanges which are to be regarded as 'communications'. They are:

- instructions
- certificates
- submissions
- proposals
- records
- acceptances
- notifications
- replies

which the contract requires.

The clause states that these should be communicated in a form which can be read, copied and recorded; which presumably means that they should be in writing.

The final sentence of clause 13.1, 'Writing is in the language of this contract', can be taken to mean that all communications are to be written in the language specified in part one of the contract data as the language of the contract.

Note that a 'decision' is not listed in clause 13.1 as a matter to be communicated. For the possible significance of this see Chapter 14, section 4.

**Clause 13.2 – receipt of communications**

Clause 13.2 states that a communication takes effect when it is received. It is not clear how this applies to certificates and the like which are dated by the sender, or how evidence of receipt can be monitored. However, the reason for making communications take effect when received is obvious enough from clause 13.3.

**Clause 13.3 – period for reply**

This clause places an obligation on the project manager, the supervisor and the contractor to reply to communications within the period of reply stated in part one of the contract data.

But note that the clause contains two exceptions to the obligation:

- it only applies to communications to which the contract requires a reply
- it does not apply if otherwise stated in the contract.

Failure by the project manager or the supervisor to comply with clause 13.3 is a compensation event (clause 60.1(6)). There is no stated sanction for failure by the contractor to comply.

The difficulty of clause 13.3 is knowing exactly what is meant by a reply which is required by the contract. Clause 13.4 expressly requires the project manager to reply to communications from the contractor for acceptances, but the wording of clause 13.3 suggests that it is intended to be of much wider effect than communications on acceptances.

**Clause 13.4 – replies on acceptances**

Clause 13.4 deals solely with communications on acceptances. These clearly do require replies within the period for reply.

Additionally the clause requires the project manager to state reasons if his reply is not acceptance and the contractor to resubmit his communication taking account of the reasons.

Matters falling within the scope of clause 13.4 include:

- clause 15.1 – proposals to add to the working areas
- clause 21.2 – particulars of the contractor's design

## 5.10 Communications

- clause 23.1 – particulars of design of equipment
- clause 24.1 – replacement persons
- clause 26.2 – names of subcontractors
- clause 26.3 – conditions of contract for subcontracts
- clause 31.1 – the first programme
- clause 32.2 – revised programmes
- clause 85.1 – insurance policies.

Each of the above named clauses has within its own text a stated reason, or reasons, for non-acceptance. Clause 13.4 provides a further general reason for withholding acceptance which applies to all the above – that more information is needed to assess the contractor's submission.

As will be seen from clause 13.8 withholding acceptance for a stated reason is not a compensation event.

There is a danger that to avoid the contractor becoming entitled to a compensation event, project managers may, if under time pressures, be disposed to ask the contractor for further superfluous information. Few project managers will want to be answerable to the employer for compensation events caused by their late replies.

### Clause 13.5 – extending the period for reply

Clause 13.5 provides for the period for reply to be extended. It is for the project manager to formally extend the period for reply but he may only do so when there is agreement with the contractor on the extension before any reply is due.

The wording of clause 13.5 suggests, but it is not completely firm on the point, that any extension of the period for reply is specific to a particular communication and is not of general effect. Since the clause seems to be principally of benefit to the employer in that an extension of the period for reply may save him from the liability of a compensation event, the incentive for the contractor to give his agreement to an extension may be correspondingly reduced. But is this, perhaps, a case where failure to give agreement might be breach by the contractor of the obligation to co-operate in clause 10.1, thereby depriving the contractor of his entitlement to a compensation event (clause 61.4) or even allowing the employer to recover any payment made under the compensation event as damages for breach?

### Clause 13.6 – issue of certificates

Clause 13.6 does no more than clarify the simple point of who is to be the recipient of certificates.

The project manager is to issue his certificates to the employer and to the contractor. The supervisor is to issue his to the project manager and to the contractor. The reason for the supervisor's certificates going to the project manager and not to the employer is the need for the project manager to have those certificates to fulfill his wider role in administering the contract.

Note that the clause applies only to certificates and not to other forms of communication.

**Clause 13.7 – notifications**

Clause 13.7 deals with notifications. It states that any notification which the contract requires is to be communicated separately from other communications. The purpose of this presumably is to avoid notifications being overlooked, or lost, in other issues. Clauses to which it applies include:

- clause 14.2 – delegation
- clause 14.4 – replacement of the project manager or supervisor
- clause 16.1 – early warning
- clause 17.1 – ambiguities and inconsistencies
- clause 19.1 – illegal or impossible requirements
- clause 40.3 – tests and inspections
- clause 42.2 – defects
- clause 61.1 – compensation events
- clause 61.3 – compensation events
- clause 61.5 – decisions on compensation events
- clause 61.6 – assumptions on compensation events
- clause 64.3 – assessment of compensation events
- clause 65.1 – implementation of compensation events
- clause 73.1 – discovery of objects of value etc.

Clause 13.7 applies to the contractor, the project manager and the supervisor.

The clause has attracted some adverse comment as a generator of paperwork and, if applied strictly, it could appear officious if not wasteful. For example, under clause 61.1 if the project manager gives an instruction he is also required to notify the contractor that a compensation event has occurred and to instruct that a quotation is submitted. It might be sensible to cover these three things in one letter but the NEC does not permit that.

**Clause 13.8 – withholding an acceptance**

The NEC takes a very direct approach to the reasons for the project manager's actions. It requires reasons to be given and it states which reasons are not compensation events.

Clause 13.8 deals with reasons for non-acceptance. It contains two distinct provisions:

- the project manager may withhold acceptance of a submission by the contractor
- withholding for a reason stated in the contract is not a compensation event

The purpose of the first provision of clause 13.8 is presumably to indicate that the project manager may withhold acceptance of a submission for any reason. That generally is the scheme of the NEC. The purpose of the second provision is not clear. The provision is simply a reflection of clause 60.1(9) which states that withholding an acceptance for a reason not stated in the contract is a compensation event. So in clause 13.8 the NEC provides that withholding for a stated reason is not a compensation event; and in clause 60.1(9) that withholding for a non-stated reason is a compensation event.

The problem with this belt and braces approach to drafting is it brings into question the effectiveness of other provisions which are not treated similarly.

For further comment on what is a reason – is it a fact, or is it an opinion – see Chapter 11, section 2 and the comment on clause 26.2 in Chapter 7, section 9.

## 5.11 The project manager and the supervisor

Clause 14 contains in its four sub-clauses a miscellaneous set of provisions relating to the project manager and the supervisor.

### Clause 14.1 – acceptance of a communication

This clause states that acceptance by either the project manager or the supervisor of a communication from the contractor or of his work does not change the contractor's responsibility to provide the works or his liability for his design. The broad intention is probably to convey the point found in most standard forms that only the parties themselves and not contract administrators or supervisors have power to change the obligations of the parties.

But it is questionable to what extent the clause is compatible with the requirements in clause 10.1 on mutual trust and co-operation. If the contractor cannot rely on acceptances given by the project manager and the supervisor, how much room is left for mutual trust and co-operation? And it is questionable whether it is fully compatible with the notion of the project manager as the employer's agent.

Note also the comment in Chapter 9, section 2 on the significance of acceptance of the contractor's design in defining what is a defect.

**Clause 14.2 – delegation**

Clause 14.2 sets out the powers and procedures for delegation by the project manager and the supervisor. As noted in section 5.6 above, there are no restrictions on what actions can be delegated or how many delegates there can be.

The difficult question of whether delegation is disposal of authority to someone else or sharing authority with someone else is not directly addressed, but the last sentence of the clause may be an indication that power sharing rather than disposal is intended.

**Clause 14.3 – instructions**

The short and apparently simple clause 14.3 is one of the most important clauses in the NEC. It states that the project manager may give an instruction to the contractor which changes the works information. The clause may not be instantly recognisable as the variation clause of the contract but that is what it is.

The simplicity of the clause must be false. The project manager cannot have an unfettered power to change the works information.

For further comment on this clause see Chapter 4, section 6.

**Clause 14.4 – replacements**

This clause entitles the employer to replace the project manager or the supervisor simply by giving notice to the contractor.

Most construction contracts have a similar provision but process and plant contracts are more restrictive on how replacements are made.

# Chapter 6

# General core clauses

## 6.1 Introduction

The common core clauses of the NEC commence in Section 1 (headed 'General') with nine sets of clauses, numbered 10 to 19, which lay the foundation for the remainder of the contract. These general clauses cover:

- actions – clause 10.1
- identified and defined terms – clauses 11.1 to 11.17
- interpretation and the law – clauses 12.1 and 12.2
- communications – clauses 13.1 to 13.8
- the project manager and the supervisor – clauses 14.1 to 14.4
- adding to the working areas – clause 15.1
- early warning – clauses 16.1 to 16.4
- ambiguities and inconsistencies – clause 17.1
- health and safety – clause 18.1
- illegal and impossible requirements – clause 19.1

Some comment is given in other chapters of this book on most of the general clauses so the commentary which follows in this chapter is for most clauses merely supplementary.

## 6.2 Actions

Clause 10.1 states broadly how the employer, the contractor, the project manager, the supervisor and the adjudicator are to act. For detailed comment on this see Chapter 5, section 3.

**Meaning of 'actions'**

What the NEC does not do, although it would undoubtedly be helpful if it did, is define what it means by an 'action'. It is, for example, fundamental to the operation of clause 90.1 on the settlement of disputes that 'action' should have a precise meaning. This is because only the contractor may refer disputes about actions of the project manager or the supervisor to adjudication. Or, put another way, the employer is not permitted to dispute actions of the project manager or the supervisor.

On its ordinary meaning an action is performing a task. The Shorter Oxford Dictionary gives its first meaning as 'the process or condition of acting or doing'. Clearly the contractor has actions to perform in providing the works, and the project manager and the supervisor have actions to perform in fulfilling their duties. The list of items to be communicated in clause 13.1 provides a typical check list of administrative actions:

- instructing
- certifying
- submitting
- proposing
- recording
- accepting
- notifying
- replying.

The question is, are disputes about the outcome of actions (opinions/ decisions) to be regarded as disputes about actions? In particular, taking perhaps the most arguable points, are disputes about the value of certificates or the assessment of compensation events truly disputes about actions? If the project manager has followed the procedures of the contract and arrived at an opinion which is different from that of the contractor, is that not a dispute about an opinion rather than a dispute about an action?

Returning then to the importance of the point as it affects clause 90.1, the position appears to be this. If a dispute about a certificate is only to be seen as a dispute about an action then only the contractor can challenge certificates. But if a distinction can be made in a dispute about a certificate between the amount on the certificate and the process of certification then either the contractor or the employer can challenge the amounts on certificates.

No doubt the matter will be settled in due course by a ruling of the courts or by amendment to the contract, but in the meantime it will remain a matter of debate, and perhaps a matter of concern to employers, as to whether the NEC does, intentionally or unintentionally, prevent the employer from disputing decisions as well as actions of the project manager.

## 6.3 Identified and defined terms

### Clause 11.1 – terms

For comment on this clause see Chapter 4, section 4.

### Clause 11.2(1) – the parties

This clause defines the term 'The Parties' as the employer and the contractor, both of whom are to be identified by name in the contract data. Although only 'the Parties' is a defined term, the NEC does use the singular 'Party' as if it is also a defined term.

For process and plant contracts where traditionally the parties are the purchaser and the contractor, and purchasers may be reluctant to contract under the title of 'employer' for various administrative reasons, a suitably drafted additional condition substituting 'purchaser' for 'employer' may be a simpler solution than amending the standard NEC conditions.

### Clause 11.2(2) – others

For comment on this clause see Chapter 5, section 2.

### Clause 11.2(3) – the contract date

For comment on this clause see Chapter 4, section 5.

### Clause 11.2(4) – to provide the works

The NEC avoids such conventional terminology as 'design, construct, complete, maintain' to describe broadly what the contractor has undertaken to do and uses instead an all embracing term 'to Provide the Works'.

Clause 11.2(4) states that to provide the works means:

- to do the work necessary to complete the works in accordance with the contract, *and*
- to do all incidental work, services and actions which the contract requires.

The defined term 'To Provide the Works' or the similar term 'Provides the Works' is found in the following clauses:

- 11.2(4)   – definition
- 11.2(5)   – constraints in the works information
- 11.2(11)  – equipment
- 14.1      – contractor's responsibility for his design
- 20.1      – contractor's main responsibility
- 23.1      – design of equipment
- 26.2      – acceptance of subcontractors
- 26.3      – terms of subcontracts
- 31.2      – programme.

In clause 15.1 (adding to the working areas) the term 'Providing the Works' is also used as a defined term.

The problem with the definition in clause 11.2(4) is that it fails to mention the contractor. So taken literally it includes by its words 'actions which this contract requires' actions to be performed by the employer, the project manager and the supervisor. Consequently an offer by the contractor in the form of tender or form of agreement 'to provide the works' is not strictly correct.

A point which is perhaps of more importance is whether the phrase 'incidental work' used in the definition can be taken as a restriction on the type of varied work which the contractor can be instructed to perform. See Chapter 7, section 1 for comment on the otherwise apparent lack of any limitations on the amount or type of variations which may be ordered under the NEC. To remedy this obviously unsatisfactory situation it may not be unreasonable to say that clause 11.2(4) distinguishes between the work necessary to complete the works (as specified at the outset) and incidental work which can be instructed by changes to the works information. This may not have been the intention of the draftsmen in using the phrase 'incidental work' but it may be more in accordance with the intentions of the parties.

### Clause 11.2(5) – works information

For comment on the clause 11.2(5) definition of works information and the general importance of the works information in the NEC see Chapter 4, section 6.

### Clause 11.2(6) – site information

For comment see Chapter 4, section 7.

### Clause 11.2(7) – the site

The NEC has two defined terms to cover areas used by the contractor in undertaking the works. The 'site' is broadly the land (or area) provided by

## 6.3 Identified and defined terms

the employer and which is identified in a drawing (or by description) in part one of the contract data. The 'working areas' are lands (or other areas) additional to the site and identified by the contractor in part two of the contract data.

The definition of the 'site' in clause 11.2(7) does little, by itself, to reveal its significance. It states only that the site is:

- the area within the boundaries of the site, *and*
- volumes above and below affected by the work.

The significance of the defined term has to be developed from the clauses which use it. In particular:

- clause 30.1 — work not to start on the site until the first possession date
- clause 33.1 — the employer is to give possession of the site to the contractor
- clause 35.1 — possession returns to the employer on take over
- clause 60.1(12) — physical conditions within the site
- clause 73.1 — objects and materials within the site
- clause 80.1 — risk of use or occupation of the site.

Thus it can be seen that the site is an area over which the employer has control which is to be made available to the contractor for the purposes of constructing the works. The employer retains certain responsibilities for the site, however, during the time it is in possession of the contractor.

The employer is required to specify the boundaries of the site in part one of the contract data. There is no obvious provision in the NEC for extending the site, which raises the question of whether it is permissible for a variation (or change in the works information) to be ordered which takes the works beyond the original site boundaries.

### Clause 11.2(8) – the working areas

The definition of working areas in clause 11.2(8) is singularly unhelpful in explaining the meaning of the term in that it states merely that the working areas are the working areas unless changed later. Nor does the fact that 'working areas' is both a defined term and an identified term greatly assist. The entry made by the contractor in part two of the contract data is simply a statement of what the working areas are in a geographical sense. Clause 15.1 (adding to the working areas) indicates only that the contractor requires the project manager's acceptance to any addition to the working areas. What then are the working areas and what is their purpose?

The answer is to be derived from clauses 70 and 71 dealing with title to, and payment for, equipment, plant and materials and in the schedules of cost components. Clause 70 suggests that the employer has title to all equipment, plant and materials within the working areas, but title outside the working areas only if marked by the supervisor. Clause 71 confirms that this is to do with payment because it states that the supervisor only marks equipment, plant and materials outside the working areas if the contract identifies them for payment.

Section 5 core clauses on payment do not deal explicitly with payment for off-site materials but there is apparently no barrier to the contractor producing an activity schedule which would trigger payments for off-site materials. And this is where the working areas take effect by giving entitlement to such payment.

Additionally, in the schedules of cost components certain distinctions are made between costs incurred within and without the working areas.

**Clause 11.2(9) – subcontractors**

For comment on the definition of a subcontractor in clause 11.2(9) see Chapter 5, section 1. For comment on subcontractors generally see Chapter 16.

**Clause 11.2(10) – plant and materials**

Clause 11.2(10) states simply that plant and materials are items intended to be included in the works.

The key word is probably 'intended' because it indicates that plant and materials do not fall within the scope of the defined term simply by being included in the works. It may well be that some plant and materials, in the ordinary sense of the words, are included in the works – perhaps as temporary works which become incorporated. But these are excluded for the purposes of the definition which is principally to do with title (clauses 70.1 and 70.2), marking for payment (clause 71.1), and the schedules of cost components.

Clearly 'plant' as mentioned in the definition does not mean contractor's plant used for constructing the works. As to precisely which plant and materials are intended to be included in the works, that is presumably to be determined from the works information.

**Clause 11.2(11) – equipment**

Equipment is defined in clause 11.2(11) as:

## 6.3 Identified and defined terms

- items provided by the contractor, *and*
- used by him to provide the works, *and*
- which the works information does not require him to include in the works.

The definition of equipment assists in distinguishing between plant and equipment, but it serves mainly in fixing a meaning to the term for the purposes of:

- clause 23.1 – design of equipment
- clauses 70.1 and 70.2 – title to equipment
- clause 72.1 – removal of equipment
- contract data entries
- schedules of cost components.

Note that although the definition refers only to 'items provided by the Contractor', clause 26.1 extends the scope to subcontractors by stating that the contract applies as if a subcontractor's equipment is the contractor's.

The wording in the last part of the clause – 'which the Works Information does not require' – is obvious enough in its meaning but it is not fully in keeping with the style of the NEC. In practice the works information is unlikely to have negative requirements other than operating restrictions.

### Clause 11.2(12) – the completion date

Clause 11.2(12) is one of the most important clauses of the NEC but like other important clauses its significance is disguised by the simplicity of its wording.

The clause states simply that the completion date is the completion date unless later changed in accordance with the contract. It may not be obvious at first reading but this is the clause which allows the time for completion to be extended. The clause works because the completion date is both a defined term and an identified term. So a completion date is entered into the contract data by either the employer or the contractor as appropriate, and that date can later be changed through the operation of the contract – which is by the compensation event mechanism.

One aspect of the wording of the clause which may attract the attention of lawyers is whether, by its reference to 'later changed', it restricts extensions of time to being granted retrospectively. If this is not the intention it is difficult to see any reason for including the word 'later'.

**Clause 11.2(13) – completion**

The meaning of 'completion' is a common source of contention in construction contracts. See, for example, the cases of *Emson Eastern Ltd* v. *EME Developments Ltd* (1991) and *H W Nevill (Sunblest) Ltd* v. *William Press & Sons* (1981).

The NEC seeks to avoid contention by defining 'completion' rather than by relying on traditional phraseology such as practical completion and substantial completion.

Clause 11.2(13) states that completion is when the contractor has:

- done all the work the works information states he is to do by the completion date, *and*
- corrected notified defects which would have prevented the employer from using the works.

But in reality this definition is no more than a description. And, subject as it inevitably is to opinion, it does not by itself fix completion. That is done under clause 30.2 where it is stated that the project manager decides the date of completion.

So in effect completion is when the project manager decides that clause 11.2(13) has been fulfilled. Whether or not this is open to dispute and review by an adjudicator or disputes tribunal is debatable. If the project manager's decision can be regarded as an 'action' then there is no problem. But if a decision is not an action then it is certainly arguable that the parties have conferred a binding power on the project manager, and his decision is not disputable.

As to the detail of the wording of clause 11.2(13) a few points are worth noting.

Firstly, the clause uses the phrase 'all the work'. 'Work' however is not a defined term and clause 11.2(4) appears to make a distinction between work necessary to complete the works and incidental work, services and actions. This could obviously lead to argument on what is required for completion and that seems to have been recognised for the clause continues 'work which the Works Information states he is to do by the Completion Date'.

Well meaning as this may be it is a surprising, if not alarming, arrangement that completion is assessed by reference to what is stated in the works information and not by a more practical test. What is the position if the works information fails to state what is to be done by completion?

The second leg of the definition – that the contractor has completed notified defects which would have prevented the employer from using the works – offers no solution to the above unless defects actually exist. It does little more than suggest that defects which do not prevent use of the works are no barrier to completion.

The retrospective phrase 'would have prevented' looks rather odd in comparison with the general style of the NEC. It seems to suggest that instead of a factual test being applied in respect of the effects of defects, a hypothetical test is applied. It is difficult to see what this is intended to achieve.

**Clause 11.2(14) – the accepted programme**

The accepted programme is one of the most important documents in the administration of an NEC contract, even though it will not always be a formally incorporated contract document. In particular, the accepted programme:

- fixes dates for possession of the site (clause 33.1)
- affects amounts due for payment (clause 50.3)
- determines the contractor's entitlements to compensation events (clauses 60.1(2), 60.1(3), 60.1(5))
- governs the assessment of compensation events (clauses 63.3 and 63.6).

The purpose of clause 11.2(14) is to define the accepted programme so that it can be identified as the programme with contractual effect. By the definition in clause 11.2(14) the accepted programme is:

- the programme identified in the contract data, *or*
- the latest programme accepted by the project manager.

The clause goes on to say that the latest programme accepted by the project manager supersedes previous accepted programmes. This appears to be a statement of the obvious but it may be significant in the assessment of compensation events. See the comment in Chapter 11, section 5.

An unusual feature of the NEC is that until the contractor has submitted a first programme for acceptance by the project manager, one quarter of amounts due in interim payments is withheld (clause 50.3).

The decision on whether or not a programme should be identified in the contract data and, as a tender programme, should formally be incorporated into the contract, is probably intended to be a decision taken by the employer. However, complications could arise if the contractor identifies a programme in part two of the contract data even when not required to do so. The risks to employers from tender programmes and method statements have been well exposed in a number of cases, most notably *Yorkshire Water Authority* v. *Sir Alfred McAlpine & Son (Northern) Ltd* (1985), and employers need to be careful with tender programmes and method statements to avoid acquiring unexpected liabilities and obligations.

**Clause 11.2(15) – defects**

For comment on this clause see Chapter 9, section 2.

**Clause 11.2(16) – the defects certificate**

Again see Chapter 9, section 2.

**Clause 11.2(17) – the fee**

The fee has two functions in the NEC:

- it applies to all the main options in the calculation of compensation events (clause 63.1), *and*
- it applies to the cost reimbursable valuations of work in main options C, D, E and F.

Clause 11.2(17) simply states that the fee is the amount (in money terms) calculated by applying the fee percentage to the actual cost. The fee percentage is an identified term stated by the contractor in part two of the contract data.

**Clause 11.2(18)**

Not used in the second edition of the NEC.

**Clauses 11.2(19) to 11.2(30)**

These clauses cover identified and defined terms for particular main options and they relate to:

- the prices
- the price for the work done to date
- actual cost
- disallowed cost.

## 6.4 Interpretation and the law

**Clause 12.1 – interpretation**

Little comment is required on clause 12.1 which is typical of the interpretation clause found in most contracts confirming that words can have singular or plural meanings and different gender applications.

**Clause 12.2 – law of the contract**

The law of the contract is an identified term to be stated in part one of the contract data.

Clause 12.2 merely confirms that the contract is governed by the law of the contract. Those concerned with international contracts, however, will understand that there may be a difference between the substantive law of the contract and the procedural law under which it is enforceable.

## 6.5 Communications

For detailed comment on clauses 13.1 to 13.8 dealing with communications see Chapter 5, section 10.

## 6.6 The project manager and the supervisor

For detailed comment on clauses 14.1 to 14.4 dealing with the project manager and the supervisor see Chapter 5, sections 6, 7, 8 and 9.

## 6.7 Adding to the working areas

The working areas are both a defined term (clause 11.2(8)) and an identified term of the NEC. Their purpose is discussed in section 6.3 above.

**Clause 15.1 – adding to working areas**

Clause 15.1 provides for adding to the working areas which have been identified by the contractor in part two of the contract data. The addition is a formal process requiring the contractor to submit his proposals to the project manager for acceptance. Reasons stated in clause 15.1 for not accepting the proposals are:

- that the addition is not necessary for providing the works, *or*
- the additional area will not be used for work in the contract.

The most obvious reason for the contractor wishing to add to the working areas is to ensure that resources used in providing the works come within the cost reimbursable scope of the schedules of cost components – either for the purposes of direct payments or the assessment of compensation events. Generally, it seems that apart from design, manufacture and fabrication, resources outside the working areas are to come within the fee.

6.7 *Adding to the working areas*

Another reason could be to permit interim payments to be made for off-site materials etc. without involving the procedures of clause 71.1 (marking equipment, plant and materials outside the working areas). But this might well involve a revision to the activity schedule unless, of course, the contractor has failed to properly identify the working areas in the first place.

This does raise the question as to whether clause 15.1 can be used to rectify an omission in the contract data on the working areas. The answer is that it can. The contract data entry reads, 'The working areas are the Site and...'. So even if the contractor leaves the entry blank the works commence with a working area corresponding to the site boundaries.

In fact if the project manager does not accept a proposal for adding to the working areas when that addition is either necessary for providing the works or the additional area will be used for work in the contract, then that is a compensation event under clause 60.1(9).

## 6.8 Early warning

One aspect of the NEC which has attracted much attention and commendation is that it provides for early warning to be given of potential problems. The idea is not new. Similar provisions have been in process and plant contracts for many years. But the NEC is far more explicit in its intentions than other contracts and additionally it puts express sanctions on the contractor for failing to comply.

**Clause 16.1 – early warning notices**

Clause 16.1 places an obligation on both the contractor and the project manager to give an early warning notice to the other as soon as either becomes aware of any matter which could:

- increase the contract price
- delay completion, *or*
- impair the performance of the works in use.

The sort of things which would obviously be included within the scope of the clause are the discovery of unforeseen ground conditions, materials shortages, design problems, insolvency of key subcontractors and the like. And the clause is so worded that it is clearly more than a mechanism for one party informing the other of its (the other's) faults. It requires confession of the parties' own faults.

Again this is not wholly new. The JCT building contracts have long

## 6.8 Early warning

required the contractor to give notice of any delay which is likely to prevent completion by the due date, whether or not the cause of the delay is a ground for an extension of time. But the NEC goes well beyond this. Note, for example, that the reference in clause 16.1 is to delay in completion not to delay in completion by the completion date.

One of the difficulties of clause 16.1, for all its good intentions, is that it suffers from the problem, common to much of the NEC, of leaving unclear how rigidly it is to be operated. Is the project manager required to give early warning before issuing any instruction or variation with cost or delay implications? Is the contractor required to give early warnings on the basis of long range weather forecasts?

The answer it may be said is to use common sense. But that brings its own difficulties. Common sense would not require the contractor to give early warning notice to the project manager that an instruction for change could increase the prices or delay completion. The contractor might expect the project manager to work that out for himself. But if it is not necessary to give early warning of matters the other party should reasonably know about, would it not be better if the clause said so?

As it stands, the parties have to make their own decisions on whether to be engulfed in early warning notices or to apply tests of reasonable knowledge or degrees of seriousness to the operation of clause 16.1.

But whatever policy is adopted on this, the parties do need to remain alert to the point that to be valid an early warning notice must be communicated separately from other communications. This follows from clause 13.7 (notifications). Consequently it may not be possible to rely on a communication such as a letter dealing with a problem, even if it is an early warning notice in all but name. Similarly the project manager will not be able to argue that an instruction by itself is effective as an early warning notice.

### Failure to give early warning notice

The sanction on the contractor for failing to give a required early warning notice is found in clause 63.4 relating to the assessment of compensation events. This clause is discussed in detail in Chapter 11, section 5. At first reading it appears to suggest that it does not matter whether the contractor gives an early warning or not. But what it presumably intends is that if the contractor does not give a required early warning, then the assessment of a compensation event cannot be greater than the assessment which would have followed the notice.

In practice the sanction may not amount to much because under clause 61.3 the contractor is required to notify compensation events within two weeks of becoming aware of them. If he fails to do so he risks losing his entitlements. So in normal circumstances it is difficult to see how the time for giving early warning can itself be more than two weeks if the contractor intends to claim a compensation event.

But in any event the sanctions on the contractor should not bite if he can show that he gave his early warning notice as soon as he became 'aware' of the matter. Curiously, though, in clause 61.5 (notifying compensation events) it is left to the project manager to decide when the contractor should have become aware.

There is no express sanction in the NEC for failure by the project manager to give early warning. But, acting in the best interests of the employer, the project manager would normally be expected to be conscientious in doing so. Any proven failure by the project manager to give early warning of a matter of which he was aware would, however, be a breach of clause 10.1 (actions) and that would arguably entitle the contractor to damages for breach against the employer or payment under the compensation event in clause 60.1(18) – breach of contract by the employer.

### Clause 16.2 – attendance at early warning meetings

In appropriate cases early warning notices will be followed by early warning meetings. Clause 16.2 allows either the project manager or the contractor to instruct the other to attend an early warning meeting.

The style of the clause is quite peremptory – instructions to attend are given. The consequences of one party finding the instructions inconvenient or impertinent and failing to attend are not addressed.

Clause 16.2 also provides that either the project manager or the contractor may instruct other people to attend if the other agrees. Taken literally, this would be dependent on the project manager or the contractor having it within their power to instruct others to attend. But what it presumably means is that if the project manager and the contractor so agree, then either can invite other persons to an early warning meeting if they think their presence would be helpful.

Note that the supervisor has no automatic right to attend and that he can only do so if the contractor agrees that he can.

### Clause 16.3 – early warning meetings

Clause 16.3 puts early warning meetings on a formal footing by requiring those who attend to co-operate in:

- making and considering proposals
- seeking solutions
- deciding actions.

## 6.8 Early warning

Clearly the obligation to co-operate only extends to those who are bound by the contract but it does nevertheless raise some interesting questions on whether it really is intended to be an obligation or merely an exhortation.

It is, of course, perfectly reasonable that those who attend early warning meetings should co-operate – so far as it is in their interest to do so. And given the obligation in clause 10.1 of the contract for the parties to act in a spirit of mutual trust and co-operation, it may be a breach of contract to be unco-operative. But there remain difficulties.

Consider, for example, a contractor who has been notified by the supervisor that some work is defective. The contractor considers that remedial works will delay completion. The contractor gives an early warning notice as he is obliged to do and instructs the project manager to attend an early warning meeting. He does not agree to the supervisor being present. The project manager, however uncharitable he may feel by being drawn into the situation, is then to co-operate in seeking a solution which will bring advantage not only to the employer but also to the contractor. So it is not open to the project manager simply to tell the contractor to get on with it and reconstruct the defective work; there would be no advantage to the contractor in that. Instead the project manager must co-operate in finding some way to improve the contractor's position.

The permutations on this theme are endless. And when it comes to defects in the contractor's design the consequences of the project manager being drawn into proposals for rectification need to be very carefully considered, not least by the project manager – with one eye on his professional indemnity insurance cover.

**Clause 16.4 – records of early warning meetings**

Clause 16.4 requires the project manager to record proposals made and decisions taken at early warning meetings. A copy of the record is to be given to the contractor.

## 6.9 Ambiguities and inconsistencies

For comment on clause 17.1 see Chapter 4, section 10.

## 6.10 Health and safety

The NEC adopts a broad approach to health and safety and avoids reference in its common clauses to specific statutes and regulations such as the Construction (Design and Management) Regulations 1994 which apply in the UK. This is in keeping with the intended overseas use of the NEC.

**Clause 18.1 – health and safety**

Clause 18.1 provides only that the contractor should act in accordance with the health and safety requirements stated in the works information. This, of course, does not in any way excuse the contractor from complying with statutory regulations if they are not mentioned in the works information; nor does it oblige the employer to set out in the works information all the health and safety requirements he believes to be applicable to the contract.

The clause should be taken as applying only to non-statutory health and safety requirements such as those particular to the employer or other interested bodies, e.g. railway regulations and the like.

## 6.11 Illegal and impossible requirements

Illegality and impossibility are difficult subjects which the NEC addresses only briefly. Points to consider are:

- whether the illegality or impossibility is absolute, such that it cannot be overcome by any means. The question then is whether the contract is totally frustrated or rendered capable of being only partially performed
- whether the illegality or impossibility is commercial or technical and is capable of being overcome by changes in the design or scope of the works
- whether the responsibility for overcoming illegality or impossibility should rest with the employer or the contractor. And that to some extent depends upon which party is responsible for design.

**Frustration**

At common law a contract is discharged and further performance excused if supervening events make the contract illegal or impossible or render its performance commercially sterile. Such discharge is known as frustration. A plea of frustration acts as a defence to a charge of breach of contract.

In order to be relied on, the events said to have caused frustration must be:

- unforeseen
- unprovided for in the contract
- outside the control of the parties
- beyond the fault of the party claiming frustration as a defence

It was said in the case of *Davis Contractors* v. *Fareham UDC* (1956) that:

> 'Frustration occurs whenever the law recognises that without default of either party a contractual obligation has become incapable of being performed because the circumstances in which performance is called for would render it a thing radically different from that which was undertaken by the contract. *Non haec in foedera veni*. It was not this that I promised to do'.

Contracts usually provide for frustration in a frustration or force majeure clause. Clause 95.5 of the NEC, which allows either party to terminate if the parties are released under the law from further performance of the whole of the contract, is perhaps to be seen as a frustration clause.

## Meaning of 'legally or physically impossible'

The meaning of the phrase 'legally or physically impossible' as used in clause 13 of the ICE Fourth edition conditions of contract was considered in the case of *Turriff Ltd* v. *Welsh National Water Development Authority* (1980).

In that case problems arose with tolerances for precast concrete sewer segments and eventually the contractor abandoned the works on the grounds of impossibility. The employer argued that impossibility meant absolute impossibility without any qualifications and since there was no absolute impossibility the contractor was in breach. The judge declined to accept that impossibility meant absolute impossibility and held that the works were impossible in an ordinary commercial sense.

## Responsibility for overcoming illegality or impossibility

At common law, in the absence of express provisions to the contrary, the employer does not warrant that the works can be built. The classic case is *Thorn* v. *London Corporation* (1876) where the contractor was to take down an old bridge and build a new one. The design prepared by the employer's engineer involved the use of caissons which turned out to be useless. The contractor completed the works with a different method and sued for his losses on the grounds that the employer warranted that the bridge could be built to the engineer's design. The House of Lords held that no such warranty could be implied.

Most contracts, however, do have provisions which indicate which party is responsible for overcoming illegality or impossibility. Usually where the employer is responsible for design, the contract requires variations to the works to be instructed so that completion can be achieved. Where the contractor is responsible for design, the burden of changing the design usually falls on the contractor.

**Clause 19.1 – illegal and impossible requirements**

Clause 19.1 requires the contractor to notify the project manager as soon as he becomes aware that the works information requires him to do anything which is illegal or impossible. The clause then states that if the project manager agrees (and by this it presumably means agrees that there is something which is illegal or impossible), the project manager is to give an instruction to change the works information appropriately.

The first point to make is that this is obviously not a frustration clause since it assumes that the contract is capable of being performed. That being the case the clause presumably applies to things which are legally or physically impossible as specified, but which can be rectified by change.

The clause does not obviously distinguish between works information provided by the employer and works information provided by the contractor, and if there are problems with either the project manager is required to give 'an instruction to change the Works Information'.

Note that the wording of the clause is not 'an instruction changing the Works Information'. Such wording would clearly be inappropriate for contractor's design where the problem was in works information provided by the contractor. But then how appropriate is it, and on what authority, can the project manager give the contractor an instruction to change the works information provided by the employer?

Such is the difficulty of trying to cover in one short clause the problems of illegality and impossibility in both employer's design and contractor's design contracts.

As to the consequence of the project manager giving his instruction to change the works information generally, it would seem that compensation event 60.1(1) will come into play.

This will not always benefit the contractor because if the instruction arises from a fault in his design (even though the fault may not be a defect within the meaning of the contract) clause 61.4 will still prevent any changes to the contract price or completion date.

# Chapter 7

# Obligations and responsibilities of the contractor

## 7.1 Introduction

The words 'obligations' and 'responsibilities' are often used as though they are synonymous. But they are not always the same or interchangeable. An obligation is a burden to be undertaken; a responsibility is a burden to be carried. Thus a contractor may have an obligation to undertake design; he may also have responsibility for design undertaken by others. The difference is obvious.

The NEC avoids the word 'obligation' but in Section 2 under the heading 'the Contractor's main responsibilities' it does detail a number of obligations. These, however, are by no means the full extent of the contractor's obligations.

### Express obligations

The style of the NEC in setting out the contractor's obligations is one of its more unusual features. Instead of using the customary word 'shall' to denote an obligation, the NEC uses present tense verbs such as 'acts', 'notifies', 'obtains'. The word 'shall' appears only in clause 10.1 (actions). But by virtue of the phrase 'shall act as stated in this contract' in that clause, it is presumed to operate through the rest of the contract.

To find the express obligations of the contractor under the NEC, therefore, it is necessary to search through the contract for present tense verbs. There are in fact 60 or so applicable to the contractor – the exact number depending upon the particular options used in any contract. They are all listed at the end of this chapter in section 7.13 to form a comprehensive schedule of express obligations.

Detailed comment in this chapter is only given on Section 2 obligations (those in clauses 20 to 29). The remaining obligations are discussed on a clause by clause basis in other chapters.

### Implied obligations

Contracts rarely attempt to detail all the obligations of the parties and even with a list of express obligations as long as that in the NEC, it is usually possible to argue that there are other obligations which should be implied to give a contract business efficacy. Some contracts do specifically exclude implied terms on particular issues, and some such as MF/1 do attempt a general exclusion of implied terms by stating that the obligations, rights and liabilities of the parties are only those as expressly stated.

The NEC has no general exclusion of implied terms which can clearly be identified as such. But it may be the case that there is very little scope for implied terms on contractor's obligations. This is because the NEC relies heavily on detail in the works information to fix obligations, and the contractor's stated obligation in clause 20.1 to 'Provide the Works' (which by definition is comprehensive of the contractor's obligations) is to do so 'in accordance with the Works Information'. This suggests that the contractor's obligations are, perhaps, limited to those in the NEC and those in the works information. If correct this leads to interesting questions on the contractual position under the NEC regarding some common obligations under other standard forms on which the NEC is silent.

**Obligations not stated in the NEC**

*To maintain confidentiality*
The NEC does not contain the provision, commonly found in process and plant contracts for good commercial reasons, that the parties must maintain secrecy and confidentiality in the contract. It may not be possible to imply such a provision so if it is required it is best to include it in the additional conditions under Option Z.

*To proceed regularly and diligently*
The NEC does not directly state any obligation for the contractor to proceed regularly and diligently (or with due expedition and without delay). Nor is there anything in the termination provisions (clause 95), dealing with breach of any such, or similar, obligations. The intention of the NEC is probably to impose discipline on the contractor through the payment scheme and delay damages, but not everyone will regard this as adequate.

*To inspect the site*
The usual clause requiring the contractor to inspect the site and to satisfy himself as to the sufficiency of his tender is missing from the NEC. However, it may be that such an obligation can be implied from clause 60.2 (compensation events) which states that, in judging physical conditions, the contractor is assumed to have taken into account information obtainable from a visual inspection of the site.

## 7.1 Introduction

*To set out the works*

There is no express obligation on the contractor in the NEC to set out the works. Perhaps it can be argued on construction works that such an obligation is implied from the definition in clause 11.2(4) of 'Provide the Works'. But this would not be particularly persuasive on process and plant contracts. To put the matter beyond doubt it is probably best in all NEC contracts to deal with setting-out obligations in the works information. This will have the added benefit of bringing problems on the accuracy of setting-out information within the scope of the compensation event procedure in so far as the project manager is required to give instructions on interpretation or accuracy of the information.

*To perform variations*

Most standard forms expressly state the contractor's obligations to perform variations. Some, such as the ICE 6th edition conditions of contract, then describe variations as changes being necessary or desirable for the completion or functioning of the works, thereby placing a sensible limit on the scope of the contractor's obligations. Others, like the IChemE forms and MF/1 have financial limits on the value of variations which may be ordered, and also prohibit variations which would prejudice the fulfilment of the contractor's obligations. The NEC does not use the word 'variations', nor the increasingly popular replacement word 'changes'. It does not even address the subject of variations directly. Instead it leads towards the contractor's obligation to perform variations by an indirect route:

- the contractor is to provide the works in accordance with the works information – clause 20.1
- the project manager may change the works information – clause 14.3
- the contractor is to obey the project manager's instructions – clause 29.1.

This, however, does not address what limitations, if any, apply to the contractor's obligations. The contractor may be able to obtain limitations by using the disputes resolutions procedures of the contract, but this would be uncertain, time consuming and unsatisfactory as a substitute for a contractual provision.

One possible argument for variations under the NEC to be of limited scope is that discussed in Chapter 6, section 3, where the meaning of the word 'incidental' in clause 11.2(4) defining the term 'Provide the Works' is discussed.

*To submit interim applications for payment*

The NEC has a detailed scheme for interim payment to the contractor but, unusually, the contractor's entitlement to payment is not dependent upon his making an application.

*To notify on completion*
The obligation to decide the date of completion is on the project manager. There is no obligation on the contractor to notify completion.

*To conform to statutes*
There is no express obligation in the NEC for the contractor to conform to statutes. This, in itself, may not be much of a problem since the absence of such a provision will not protect the contractor from unlawful actions. However, in most standard forms the obligation to conform to statutes is accompanied by an obligation to indemnify the employer against claims etc. arising from any breach – and this is potentially a serious omission from the NEC.

## 7.2 Design obligations, responsibilities and liabilities

As a matter of policy the NEC seeks to achieve maximum flexibility on the allocation of design obligations between the parties. The intention is that its standard provisions should be applicable to the entire range of situations between no contractor's design and full contractor's design. For any particular contract the extent of the contractor's design obligations is to be determined from the works information.

This is more ambitious than it might at first appear. Other standard forms of contract operate satisfactorily with flexibility in the extent of contractor's design, so why not the NEC? The difference is that the other standard forms are, in the main, drafted for either employer's design or contractor's design and any flexibility operates within a framework of recognised rules for that type of contract. For example, the ICE 6th edition conditions of contract are essentially employer's design and the supervisory role of the engineer in the contract reflects this. The IChemE model forms are essentially contractor's design and the provisions on variations reflect this. The NEC, however, is drafted to be neutral between the two situations and that presupposes that it is possible for the various provisions of the contract to operate equally well in any situation.

But the potential pitfalls of this approach are only too apparent. Consider, for example, clause 60.3 on unexpected physical conditions. That clause states that if there is inconsistency in the site information the contractor is assumed to have taken into account the physical conditions more favourable to doing the work. Perfectly reasonable, if not perhaps a little generous to the contractor, for employer's design and where the

## 7.2 Design obligations, responsibilities and liabilities

contractor is simply building the works. But apply the clause to contractor's design and what then are the consequences? Surely the essence of sound design is that it should cater for the worst conditions which might be expected not the best.

### Key question on design

Quite apart from the general question raised above on whether a contract can satisfactorily be wholly flexible on design, there are three key questions which apply to all contracts in which there is some element of design. These are:

- How is the obligation to undertake the design allocated between the parties?
- How is the responsibility for the effectiveness of the design to be allocated between the parties?
- What standard of liability attaches to the party responsible for design – is it fitness for purpose or the use of reasonable skill and care?

### Design obligations under the NEC

Clause 21.1 of the NEC deals with the extent of the contractor's design obligations. The contractor is required to design the parts of the works which the works information states he is to design. This may seem reasonably clear but see the comment below in section 7.4 on whether a design obligation can be imposed by an instruction changing the works information.

### Design responsibility under the NEC

The intention of the NEC appears to be that each party is generally responsible for the design which it undertakes. That appears to follow from the definition of defect in clause 11.2(15) and elsewhere in the contract.

There is certainly no provision in the NEC of the type found in the ICE Design and Construct conditions of contract and in MF/1, that the contractor is responsible for all the design including that undertaken by or on behalf of the employer. So the NEC will only provide single point responsibility on design when the contractor is given the obligation to undertake all the design.

One possible area of uncertainty on design responsibility is discussed in Chapter 9, section 2 under defects. That arises out of the reference in clause 11.2(15) to the contractor's design accepted by the project manager. Its effect is arguably to transfer design responsibility to the employer.

**Design liability under the NEC**

It is generally accepted that in the absence of express provisions to the contrary, there is an implied term in design and build contracts that the finished works will be reasonably fit for their intended purpose (*IBA* v. *EMI* (1980)). The significance of this is that it puts the contractor's liability for his design on a different legal basis from that of a professional designer. The law does not normally imply terms of fitness for purpose into contracts for the supply of professional services. See, for example, *Greaves Contractors Ltd* v. *Baynham Meikle & Partners* (1975) and many other cases. The duty of a professional designer is to use reasonable skill and care.

In the event of a design failure, the difference between the two standards of liability can be critical to the position of the parties. If the contractor's liability is fitness for purpose that is a strict liability and the test for liability is whether the works are fit for their intended purpose. The contractor may have used all reasonable skill and care in his design but if the specified contractual objective is not achieved, the contractor will be liable to the employer for damages. However, if the contractor's liability is limited to skill and care corresponding to that of a professional designer, then negligence must be proved to establish breach of duty. And even though design failure may occur, the contractor will not be liable if he has used reasonable skill and care.

To relieve the contractor of the higher standard of liability imposed by fitness for purpose, some design and build contracts (for example, JCT 1981) limit the standard of the contractor's design liability to the same as that of a professional designer.

The NEC in its core clauses does not have any such limitation of liability and the contractor's liability for his design is almost certainly on a fitness for purpose basis. However, Option M does purport to limit the contractor's liability to reasonable skill and care, although it is arguable that on proper interpretation of its wording it does not actually achieve this. See Chapter 3, section 7 for further comment.

There is one limitation of liability for contractor's design in the core clauses of the NEC. But that limitation, in clause 21.5, applies to the amount of damages for latent defects and it is not a limitation of the standard of liability.

## 7.3 Providing the works

### Clause 20.1 – obligation to provide the works

The NEC avoids the usual lengthy statement of the contractor's general obligations and responsibilities and relies instead on the single short sentence in clause 20.1 stating that the contractor provides the works in accordance with the works information. The key to the effectiveness of this clause lies in the two defined terms 'Provide the Works' and 'Works Information'.

The term 'Provide the Works' is defined in clause 11.2(4). It covers both the obligation to complete the works and the obligation to provide whatever is required by the contract. The term 'Works Information' is defined in clause 11.2(5). It specifies and describes the works, and details any constraints on the contractor.

Both defined terms are discussed in some detail in Chapter 6 but the point to be made here (as so often elsewhere in this book) is that the extent of the contractor's obligations is reliant on the requirements in the works information. This follows from the simplicity of the wording of clause 20.1, and in particular the words 'in accordance with the Works Information'.

Were it not for the power given to the project manager to change the works information (clause 14.3), the contractor's obligations under clause 20.1 would be inadequate in the absence of an express obligation to perform variations – which the NEC lacks.

Clause 20.1, therefore, has to be seen not only as stating the contractor's obligations to complete the works as originally defined, but also as the starting point of the contractor's obligation to perform variations.

Note that, by itself, clause 20.1 does not address in any way the matter of the contractor's entitlement to payment for the performance of his obligations. The clause is not a statement of the type found in some other contracts that the contractor's price includes all things necessary to fulfil his obligations. The contractor's entitlement to payment is dependent upon which of the main options is used for the particular contract.

### Clause 20.2 – management obligations

This clause applies only in Option F – the management contract. The purpose of the clause is to identify which parts of the works the contractor shall subcontract and which parts he may or may not subcontract. It is unfortunate that the wording of the clause is ambiguous on its key issue.

The intention of clause 20.2 is probably as follows:

- the contractor manages his own design and the construction and installation of the works (these things he may not subcontract)

- the contractor subcontracts such design, construction, installation and other work as the works information requires to be subcontracted (these things he must subcontract)
- the contractor may either subcontract or perform himself design, construction, installation and such other work as is not required to be subcontracted (these things he may subcontract).

The problem with the wording of clause 20.2 is the way the phrase 'other work' is used in the second and third sentences. It leads to an alternative from that given above as the probable intention, namely that the contractor must subcontract all design, construction and installation and must also subcontract such 'other work' as stated in the works information.

If this latter interpretation is wrong and the first is correct, then the clause leaves open how design, construction and installation which is not required by the works information to be subcontracted is actually to be performed.

### Clause 20.3 – practical implications of design and subcontracting

This clause applies in Options C, D, E and F. All are contracts which are to some extent cost reimbursable and which assume a close degree of co-operation and openness between the parties.

Clause 20.3 states that the contractor advises the project manager on the practical implications of the design of the works and on subcontracting arrangements. Both obligations are aimed at achieving cost and practical efficiency and are normal for the type of contract.

### Clause 20.4 – forecasts of total actual cost

Clause 20.4, like clause 20.3, applies in Options C, D, E and F. Again it is related to the cost reimbursable nature of the contracts.

The clause requires the contractor to prepare regular forecasts of the total actual cost at intervals as stated in the contract data for submission to the project manager. Such forecasts are required from the starting date until completion of the whole of the works. With each forecast the contractor is required to explain any changes from the previous forecast. The forecasts are to be prepared in consultation with the project manager.

There is no clear contractual purpose for the forecasts and they appear to be more by way of good management practice in financial awareness and cost control.

## 7.4 The contractor's design

### Clause 21.1 – contractor's design

This clause expresses the contractor's obligations for design. It states simply that the contractor designs the parts of the works which the works information states he is to design. Note that the clause deals solely with the obligation to design and that it says nothing about responsibility for design or the standard of liability.

It could be argued from the phrase 'parts of the works' that design by the contractor of the whole of the works is not intended. However, since all the parts constitute the whole it is not thought that the argument has much weight.

Of more concern is the argument as to whether design obligations can be imposed on the contractor by instructions from the project manager changing the works information. In many cases such an imposition (if it exists) could be wholly unreasonable and impracticable. It could change the very basis of the obligations which the contractor had expected to perform.

Such considerations are themselves good arguments against the imposition of design obligations by instruction. A further argument is that clause 21.1 refers to design which the works information 'states'. And that, it can be said, would not extend to a statement in an instruction changing the works information.

One aspect of clause 21.1 which compilers of NEC contract documents should note with care is that it probably excludes any implication that design obligations fall on the contractor because common sense interpretation of drawings or instructions suggests that they should. It is essential that the works information should state all design obligations and that nothing should be left to chance. See *Shanks & McEwan (Contractors) Ltd v. Strathclyde Regional Council* (1994) for a case under the ICE 5th edition conditions of contract on the design of specialised components.

### Clause 21.2 – acceptance of the contractor's design

Clause 21.2 deals with the acceptance of particulars of the contractor's design. The clause has three principal elements:

- the contractor's obligation to submit particulars of his design
- reasons for not accepting the design
- prohibition on the contractor from proceeding until the design is accepted.

Although clause 21.2 is not expressly limited to design of the permanent works as opposed to the design of the temporary works, that limitation can be deduced from clause 23.1 which deals with temporary works.

The obligation on the contractor to submit particulars of his design is expressly linked in clause 21.2 to requirements in the works information. In the absence of such requirements the contractor would be entitled to proceed without submitting any particulars for acceptance. That could create difficulties for the project manager in some circumstances but the project manager has the power to remedy the situation by changing the works information (clause 14.3). Such a change would then be a compensation event 60.1(1).

Note that the clause deals with 'particulars' of the design and not with the contractor's design proposals in a broader sense. Approval of such proposals is a pre-contract function which is formally recognised by the acceptance of the works information provided by the contractor in part two of the contract data and its incorporation into the contract. In the not uncommon event of a dispute arising during construction as to whether the contractor's proposals do fully meet the employer's requirements, that is dealt with under the NEC by clause 17.1 (ambiguities and discrepancies) and clause 63.7 (assessing compensation events).

The NEC does not specify the time for submission of design particulars in any of its clauses or in the standard contract data entries. However in some contracts, for obvious practical reasons, the timescale for the submission of design particulars is important – particularly if there is a technical link with other contracts. In some contracts liquidated damages are specified for late submission of design particulars. In the NEC if discipline of this order is required it will have to be introduced through specifications or requirements in the works information, contract data or additional conditions of contract.

The NEC does, however, impose a timescale for the project manager to reply to the contractor's submission (clauses 13.3 and 13.4) with the period of reply stated in the contract data. Failure by the project manager to reply in time is a compensation event – clause 60.1(6).

Clause 13.4 requires the project manager to state reasons if his reply to any submission is not an acceptance. The project manager is not restricted in his reasons (clause 13.8), but any reason which is not one stated in the contract is a compensation event – clause 60.1(9).

The only reasons stated in clause 21.2 for not accepting the contractor's design particulars are:

- non-compliance with the works information
- non-compliance with the applicable law.

For comment on the implications of acceptance of the contractor's design by the project manager see Chapter 9, section 2.

## 7.4 The contractor's design

The prohibition in clause 21.2 on the contractor proceeding with relevant work until his design is accepted is sensible for a variety of reasons – most obviously the avoidance of abortive work.

### Clause 21.3 – submission of design in parts

Clause 21.3 states that the contractor may submit his design for acceptance in parts if the design of each part can be fully assessed.

It would be odd if the contractual requirement was otherwise. One of the attractions of contractor's design is the potential for a quick start with the design being finalised as the works proceed.

### Clause 21.4 – indemnity against claims

Clause 21.4 requires the contractor to indemnify the employer against claims, compensation and costs due to the contractor infringing patents or copyrights.

Provisions of this type are included in most standard forms. Usually they state the scope of the indemnity in more detail than in the NEC and they include a counter indemnity by the employer in respect of his design, specification or use of the works.

The scope of clause 21.4 is not wholly clear. It arguably relates only to the contractor's design, on the basis that the side note to the clause is of contractual effect and that the location of the clause in the contract is relevant. If this is the case clause 21.4 is not strictly comparable with the patent indemnity clauses of other contracts which cover materials, plant and equipment as well as design.

One argument for restricting the indemnity under clause 21.4 to contractor's design is that to give it any wider scope could involve the contractor indemnifying the employer against patent infringements resulting from compliance with the employer's requirements in the works information. Most standard forms make it clear that the contractor's indemnity does not extend beyond matters of the contractor's own choice, but taking clause 21.4 literally the contractor's indemnity under the NEC is unqualified.

### Clause 21.5 – limitation of liability

Clause 21.5 allows for a financial limit to be placed on the contractor's liability for defects in his design. However, the clause only operates when an amount is stated in the contract data.

The limitation, when it does operate, applies only to defects due to the contractor's design not listed on the defects certificate. The intention of this is that the limitation should only apply to defects which are latent at the defects date and not as a general limitation of liability. Moreover the limitation in clause 21.5 is expressly stated to be an amount which is in addition to any liquidated damages due for delay or low performance.

One practical point which may need to be considered, particularly when using the NEC for process and plant contracts, is that the standard arrangement in the NEC is for the amount of any limit to be stated by the employer in part one of the contract data. However, in the contract negotiation process it may well be that it is the contractor rather than the employer who is concerned with fixing the amount of the limit.

Note that the limitation in clause 21.5 is wholly separate from, and quite different in its nature from, the limitation of liability in Option M on reasonable skill and care.

However, there is one aspect of both limitations which may come in for some scrutiny in the years ahead. That is whether the contractor's 'design' as mentioned in the NEC is inclusive of plant and materials selected by the contractor. In the ordinary course of things that would usually be the case, but with the NEC restriction of the contractor's 'design' to what is stated in the works information it may not always be so.

## 7.5 Using the contractor's design

### Clause 22.1 – employer's use of the contractor's design

Clause 22.1 deals with the employer's right to use the contractor's design for certain purposes. Under clause 22.1 the employer may use and copy the contractor's design for any purpose which is either:

- connected with the construction, use, alteration or demolition of the works (unless otherwise stated in the works information), *or*
- is stated in the works information.

This is a complex legal subject which can easily give rise to disputes on copyright and patents. The standard process and plant forms are significantly more detailed than the NEC and express the employer's rights in terms of restricted licences. A question, for example, which might be asked of the NEC is, whether the phrase in clause 22.1 'any purpose connected with construction, use, alteration, or demolition' cover extensions of the works.

The answer to that question, it is suggested, is probably not but it does indicate the care which needs to be taken by both parties in any statement they make in the contract data relating to use of the contractor's design.

## 7.6 Design of equipment

Equipment is defined in clause 11.1(11) as items provided by the contractor and used by him to provide the works but which the works information does not require him to include in the works.

In conventional contracts such equipment would usually come within the scope of either 'temporary works' or 'contractor's equipment'.

**Clause 23.1 – design of equipment**

Clause 23.1 states the obligation of the contractor to submit particulars of the design of equipment for acceptance, but only when instructed to do so by the project manager. It also states the reasons for not accepting the design which do not qualify as compensation events. These reasons are that the design will not allow the contractor to provide the works in accordance with:

- the works information
- the accepted contractor's design, *or*
- the applicable law.

There are some similarities of procedure between clause 23.1 and clause 21.2 (submission of design of parts of the works) in that the project manager can reject a design for any reason. However, there are also some notable differences.

The first difference is that clause 23.1 is activated by a project manager's instruction, and that is a communication which imposes on the contractor the obligation to reply within the stated period for reply. The second difference is that there is no prohibition in clause 23.1 on the contractor proceeding with his work until acceptance of the design is received.

This latter point is one of considerable practical importance. It emphasises that the contractor can proceed at his own risk but if he then receives a notice of non-acceptance from the project manager he is obliged by clause 13.4 to resubmit his design taking into account the reasons given for non-acceptance.

## 7.7 People

The NEC, as part of its good management strategy, places considerable importance on the quality of key staff employed by the contractor. Firstly, the contractor is required to state in part two of the contract data the names, qualifications and experience of key people. Clause 24 then deals with the contractor's obligations in respect of key people and other employees.

## Clause 24.1 – key persons

The first obligation stated in clause 24.1 is that the contractor should employ each key person named in the contract data to do the job described for him or should employ a replacement person accepted by the project manager.

It is up to the employer in the instructions to tenderers to specify how many key persons and for what jobs entries are to be made in the contract data. The numbers will obviously depend on the complexity of the project.

Failure by the contractor to employ either named key people or accepted replacements is a breach of contract on the part of the contractor. But, as is often the case with the contractor's breaches of procedural regulations, the employer's remedies are non too apparent. In serious cases the project manager could arguably give an instruction suspending work under clause 34.1 until the breach was remedied – relying on clause 61.4 to prevent the contractor recovering cost and delay as a compensation event. In the extreme the contractor's breach might be classed as a default under clause 95.2 entitling the employer to terminate.

The second obligation in clause 24.1 is that the contractor shall submit the name, qualifications and experience of each proposed replacement person to the project manager for acceptance. This obligation it is suggested applies only in respect of named key persons and not to other key persons employed by the contractor who are not named in the contract data.

The only reason stated in clause 24.1 for the project manager not to accept a proposed replacement is that his qualifications and experience are not as good as those of the person being replaced. Non-acceptance for any other reason is a compensation event (clause 13.8).

It is right that the contractor should not be unduly restricted in who he employs as key persons because regard must be had to such practical matters as staff changes and other workload commitments.

## Clause 24.2 – removal of an employee

Clause 24.2 is of wider application than clause 24.1 in that it refers to 'employees' rather than to 'key persons'.

The clause gives the project manager the power to instruct the contractor to remove an employee – but only after having stated his reasons for the instruction. The contractor is obliged to respond with speed and from one day after the instruction the employee is to have no further connection with the work included in the contract.

The project manager's power here is quite draconian and the clause goes far beyond the usual clause in standard forms relating to the removal of workpeople from site for misconduct or negligence. Firstly, there is no stated restriction on the project manager's reasons. However, unreasonable use by the project manager of his powers would be a breach of clause 10.1 (obligation to act in a spirit of mutual trust and co-operation) and the contractor might then be able to argue for a compensation event under clause 60.1.

Secondly, and perhaps of more concern, is the timescale for the removal of the employee and the prohibition on his future connection with the contract. Finding a suitable replacement within a day seems a difficult enough burden for the contractor, but banning the employee from work connected with the contract – not merely removing him from the site – seems to be going too far.

Any project manager considering exercising his powers under clause 24.2 should give some thought to potential legal liabilities to the employee concerned if his livelihood is unreasonably put in jeopardy by the project manager's actions.

## 7.8 Co-operation

Clause 25.1 is a brief clause but its implications are extensive. Amongst other things it is effectively a statement that the contractor is not entitled to exclusive possession of the site. But it is questionable whether the clause achieves another purpose, if indeed such a purpose is intended: that the contractor should provide facilities for others on the site.

### Clause 25.1 – co-operation with others

The clause states simply two obligations of the contractor:

- to co-operate with 'Others' in obtaining and providing information they need in connection with the works
- to share the working areas with 'Others' as stated in the works information

In most standard forms there are express obligations on the contractor to co-operate with, and to provide reasonable facilities to, others who are lawfully on the site. Such provisions are necessary for practical reasons and they reduce the contractor's grounds of claim in respect of obstruction by others.

In the NEC only clause 25.1 appears to come close to dealing with these matters. That suggests that the purpose of clause 25.1 is to oblige the contractor to co-operate with and provide reasonable facilities for others. But, on examination of its wording, it is doubtful if that can be read into the clause.

## 7.8 Co-operation

The clause is quite specific in referring to co-operation in obtaining and providing information. It says nothing about facilities or working arrangements. In fact, taken literally, the clause imposes on the contractor an unusual obligation to act as a co-ordinator and provider of information. It is not clear how (if at all) the contractor gets paid for this task or what responsibility he carries for the information provided.

Note that there is no restriction on the obligation in the clause to obtain and provide information to others by reference to what is stated in the works information. Consequently it may be difficult for the contractor to argue that the obligation, however onerous it becomes, is ever a compensation event.

The obligation in clause 25.1 for the contractor to share the working areas with others is, by contrast, expressly related to what is stated in the works information. The question, of course, is what level of detail will be found in the works information in respect of this. If it is just a list of the bodies who will at some time or other be on site, that is one thing; but if it is a schedule of the duration times of their activities with planned starting and completion dates, then the scope for compensation events under clause 60.1(5) is greatly increased.

In the event that for practical reasons the contractor has to share the site with others not stated in the works information, that might just come within the scope of compensation events – clause 60.1(5). But the contractor might have a more secure case for any claim if the project manager can be persuaded to change the works information by instruction.

## 7.9 Subcontracting

The provisions in the NEC for the regulation of subcontracting are strict and detailed. Strict, because appointment of a subcontractor for substantial work before acceptance by the project manager is expressly made grounds for termination of the contract (clause 95.2). Detailed, because the provisions extend to control over the terms of subcontracts.

This degree of regulation goes against the trend of other recent ICE forms of contract where the contractor's freedom to subcontract as he thinks fit has been introduced as being in keeping with modern commercial practice. However, the NEC is designed to suit a variety of procurement options, some of which are cost reimbursable, and it is normal to have strict controls on subcontracting in such contracts.

**Nominated subcontracting**

The NEC does not provide expressly for nominated subcontracting, so contractually all subcontractors are to be regarded as domestic subcontractors whether or not they are named or otherwise fixed by the requirements of the works information.

This can create a problem for the main contractor if the terms of business of the named subcontractor cannot be brought into line with those of the main contract. In such circumstances it is suggested that the contractor should seek instructions from the project manager so that changes are made in the works information to accommodate the situation.

**Assignment**

The second edition of the NEC omits the prohibition of assignment of benefits which was in the first edition. Such a prohibition could, of course, be reinstated in the second edition by the inclusion of an additional condition. But since the advantage of such a prohibition inclines towards the contractor (in that it may eliminate his liabilities for latent damage if the works pass into new ownership), there is not much incentive for the employer to include such an additional condition.

**Clause 26.1 – responsibility for subcontractors**

Clause 26.1 has two provisions. The first confirms the contractor's responsibility under the contract for work which is subcontracted. It does this by stating that the contractor is responsible for performing the contract as if he had not subcontracted.

A similar provision is found in most standard forms but it is probably more of a contractual safeguard than of strict legal necessity. Only if there are words in the contract suggesting that the contractor is entitled to some contractual relief for the defaults of his subcontractors, can the contractor claim such relief.

The second provision of clause 26.1 states that the contract applies as if a subcontractor's employees and equipment are the contractor's. This is potentially more complex in contractual and legal terms than the first provision. For example, it would seem to give the project manager the right to impose the provisions of clause 24 (people) on subcontractors. And it may have implications on the extent of the contractor's liability for the negligence of a subcontractor's employee.

**Clause 26.2 – acceptance of subcontractors**

Clause 26.2 deals with the acceptance of subcontractors. The contractor is required to submit the name of each proposed subcontractor for acceptance. And the contractor is not permitted to appoint a subcontractor until the project manager has accepted him.

The clause does not detail the criteria to be used by the project manager in examining the subcontractor's credentials (e.g. operates a QA system, has suitable experience, is financially sound) but it merely states that a reason for not accepting a subcontractor is that his appointment will not allow the contractor to provide the works.

This will not often be a straightforward matter of fact. More often than not it will be a matter of opinion. Why should the contractor propose a subcontractor who will not allow him to provide the works? So unless the reason for non-acceptance is simply that the proposed subcontractor is not the firm named in the works information and, therefore, as a fact the appointment will not allow the contractor to provide the works in accordance with the contract, or some such similar clear circumstances apply, there is likely to be a difference of opinion between the contractor and the project manager on the matter.

Unless this is resolved by agreement the adjudicator will have to be brought in. But for practical and administrative reasons the contractor may already have been obliged to propose (and appoint) an alternative subcontractor. In such a case the task of the adjudicator would then be to determine if the contractor was entitled to a compensation event for the non-acceptance.

Of course, if the project manager's reason for non-acceptance of a subcontractor is other than the stated reason of not allowing the contractor to provide the works, the contractor is entitled to a compensation event under clause 60.1(9).

One aspect of clause 26.2 which may cause some concern is that there is no express exception from the acceptance procedure for minor subcontracts or subcontracts for the supply of plant and labour only.

Because of its firm rules on the timescale for replies to submissions and other communications the NEC should avoid the delay problems which frequently accompany subcontractor acceptances under other contracts. And, unlike other contracts, the NEC has an express remedy (compensation event 60.1(6)) for any delay caused.

**Clause 26.3 – conditions of subcontracts**

Clause 26.3 deals with the conditions of contract for subcontracts. Instead of such conditions being left solely as a commercial matter for the contractor to decide – as is the case with most standard forms (except for certain cost-reimbursable contracts) – the NEC imposes a measure of control on the conditions.

The contractor is required to submit the proposed conditions of contract for each subcontract to the project manager for acceptance unless:

- the NEC subcontract or the NEC professional services contract is to be used, *or*
- the project manager has agreed that no submission is required.

## 7.9 Subcontracting

This latter point seems to be a matter for the project manager's absolute discretion.

Clause 26.3 further provides that the contractor shall not appoint a subcontractor until the project manager has approved the conditions, and the clause states as the reasons for non-acceptance:

- the conditions will not allow the contractor to provide the works, *or*
- the conditions do not include a statement that the parties to the subcontract shall act in a spirit of mutual trust and co-operation.

As discussed above there is much scope for dispute on what will, or will not, allow the contractor to provide the works. But subcontractors should welcome the employer's concern that subcontracts should include for mutual trust and co-operation.

However, to the extent that the employer might be seen in law to be taking on some responsibility for the terms of subcontracts, the project manager in exercising his discretion on whether or not to concern himself with the conditions of subcontracts may have to be careful that he does not leave the employer exposed to liability to subcontractors for unfair subcontracts.

### Clause 26.4 – contract data for subcontracts

Clause 26.4 applies only in main options C, D, E and F. It extends the control of the project manager over subcontracts so that he is entitled to examine the contract data of proposed subcontracts.

The clause applies when:

- the NEC subcontract or NEC professional services contract is used, *or*
- when the project manager instructs the contractor to make a submission.

The only stated reason for non-acceptance is that the proposed contract data will not allow the contractor to provide the works.

The purpose of this clause is essentially to protect the employer from wasted expenditure on the cost reimbursable elements of the contracts involved.

## 7.10 Approval from others

### Clause 27.1 – approval from others

Clause 27.1 is one of the shortest clauses in the NEC but its brevity belies its significance. The clause simply states that the contractor obtains approval of his design from 'Others' where necessary.

The danger of the clause is that it does not obviously reveal its intention or its scope. In particular, note that it does not state that the contractor obtains approvals as required by the works information. In other words the clause operates differently from much of the NEC. However, in doing so it leaves the contractor with the responsibility of obtaining all necessary approvals for his design which are required by statute or otherwise.

Contractors taking on design obligations under the NEC need to exercise the greatest caution, therefore, as to what approvals are required. The most obvious example is planning consent.

In the event of problems on approvals developing, various clauses of the NEC could come into play. Thus, delay would invoke clause 16.1 on early warning; illegality would invoke clause 19.1.

## 7.11 Access to the work

### Clause 28.1 – access to the work

This clause states the obligation of the contractor to provide access to the project manager, the supervisor, and others notified by the project manager, to work being done and to plant and materials.

The clause is poorly worded so that it appears to apply to plant and materials being stored for the project managers and others. However, it is almost certainly intended to apply to the work of, and the goods and materials of, the contractor and subcontractors whether on or off the site.

The contractor's obligations are not stated in terms of requirements in the works information and the contractor may have difficulty in recovering costs of requirements he considers to be excessive. Disputes on this can easily develop. For example, does the phrase 'provides access' extend to an obligation on the contractor to provide telescopic lifting platforms for the supervisor if the supervisor deems it necessary, or does it simply mean that the contractor shall allow the supervisor to use access facilities already in place or being used by the contractor.

## 7.12 Instructions

### Clause 29.1 – instructions

Clause 29.1 is one of the most important provisions in the NEC. It requires the contractor to obey an instruction which is given to him by the project manager or the supervisor and which is in accordance with the contract.

By complying with the clause the contractor takes on obligations, but under other clauses he then acquires entitlements. And providing the instructions do not result from his defaults, the contractor will usually be able to claim a compensation event under one or more parts of clause 60.

One difficult aspect of clause 29.1 is what is the position if the contractor considers that an instruction is not in accordance with the contract? Can he then refuse to comply? The answer from clause 90.2 (settlement of disputes) would appear to be no. That clause requires the parties to proceed as though a disputed matter was not disputed until resolved by the adjudicator.

## 7.13 Express obligations of the contractor

**Core clauses**

| | |
|---|---|
| 10.1 | – to act as stated in the contract and in a spirit of mutual trust and co-operation |
| 13.1 | – to communicate in a form which can be read, copied and recorded |
| 13.3 | – to reply to a communication within the period for reply |
| 13.4 | – to re-submit a communication which is not accepted within the period for reply |
| 13.7 | – to communicate notifications separately from other communications |
| 16.1 | – to give early warning of matters with delay, cost or performance implications |
| 16.3 | – to co-operate at early warning meetings |
| 17.1 | – to give notice of ambiguities or inconsistencies in the documents |
| 18.1 | – to act in accordance with health and safety requirements |
| 19.1 | – to give notice of any illegality or impossibility in the works information |
| 20.1 | – to provide the works in accordance with the works information |
| 21.1 | – to design such parts of the works as stated in the works information |
| 21.2 | – to submit particulars of his design for acceptance |
| 21.4 | – to indemnify the employer against claims for infringements of patents or copyrights in the contractor's design |
| 23.1 | – to submit when instructed particulars of design of items of equipment |
| 24.1 | – to employ key persons as stated in the contract data or acceptable replacements |
| | – to submit relevant details of proposed replacements |
| 24.2 | – to remove any employee on the project manager's instructions |
| 25.1 | – to co-operate with others in obtaining and providing information |
| | – to share the working areas with others as stated in the works information |

## 7.13 Express obligations of the contractor

- 26.2 – to submit the names of proposed subcontractors for acceptance
- 26.3 – to submit the proposed conditions of contract for each sub-contract for acceptance
- 27.1 – to obtain approval of his own design from others where necessary
- 28.1 – to provide access to the works to the project manager, supervisor and others
- 29.1 – to obey instructions given by the project manager or the supervisor which are in accordance with the contract
- 30.1 – to do the work so that completion is on or before the completion date
- 31.1 – to submit a programme for acceptance within a period stated in the contract data
- 31.2 – to show various details in each programme
- 32.1 – to show various details in revised programmes
- 32.2 – to submit a revised programme when instructed to or as required in the contract data
- 33.2 – to provide facilities and services as stated in the works information
  – to pay the assessed cost of not providing facilities and services
- 36.2 – to submit a quotation for acceleration when so instructed
- 40.2 – to provide materials, facilities and samples for tests and inspections as stated in the works information
- 40.3 – to notify the supervisor of tests and inspections before they start
  – to notify the supervisor of the results of tests and inspections
  – to notify the supervisor before doing work which would obstruct tests or inspections
- 40.4 – to correct defects revealed by tests or inspections and to repeat such tests or inspections
- 40.6 – to pay the assessed cost incurred by the employer in repeating tests or inspections
- 42.1 – to carry out searches as instructed by the supervisor
- 42.2 – to notify the supervisor of defects found before the defects date
- 43.1 – to correct defects
  – to correct notified defects before the end of the defects correction period
- 45.1 – to pay the assessed costs of having defects not corrected within the defects correction period corrected by others
- 51.1 – to pay the employer if an interim assessment reduces the amount due from that already paid
- 61.1 – to put instructions or changed decisions into effect
- 61.3 – to give notice of a compensation event
- 61.4 – to submit instructions for compensation events if instructed to do so
- 62.1 – to submit alternative quotations for compensation events if instructed to do so

## 7.13 Express obligations of the contractor

62.3 – to submit quotations for compensation events within 3 weeks of being instructed to do so
62.4 – to submit revised quotations for compensation events within 3 weeks of being instructed to do so
72.1 – to remove equipment from the site when it is no longer needed
73.1 – to notify the finding of any object of value, historical or other interest
81.1 – to carry risks which are not the employer's risk from the starting date until the defects certificate is issued
82.1 – to make good loss or damage to the works until the defects certificate is issued
83.1 – to indemnify the employer against claims due to contractor's risks
84.1 – to provide insurances as required by the contract
85.1 – to submit insurance policies and certificates for acceptance
85.3 – to comply with the terms and conditions of insurance policies
86.1 – to pay the costs incurred by the employer in covering insurances which are the contractor's responsibility
87.1 – to accept insurance policies and certificates taken out by the employer if they comply with the contract
90.2 – to proceed with matters in dispute which are referred to adjudication as though they were not disputed until there is a settlement
91.1 – to provide information to the adjudicator
96.2 – to leave the working areas and remove equipment on termination.

## Main option clauses

### Option A

31.4 – to show the start and finish of activities on the activity schedule on each programme
54.2 – to submit revisions to the activity schedule so that it is compatible with the accepted programme.

### Option B  Nil

### Option C, Option D, Option E

20.3 – to advise the project manager on the practical implications of the design of the works and on subcontracting arrangements
20.4 – to prepare forecasts of the total actual cost for the whole of the works

26.4 – to submit the proposed contract data for each subcontract for acceptance
31.4 – to show the start and finish of each activity on the activity schedule on each programme (Option C only)
36.5 – to submit a subcontractor's proposal to accelerate for acceptance
52.2 – to keep records of costs and payments
52.3 – to allow the project manager to inspect accounts and records.

## Option F

20.2 – to manage the contractor's design and the construction and installation of the works
 – to subcontract such design, construction and installation as is stated in the works information to be subcontracted
 – to do work not stated in the works information to be subcontracted himself or to subcontract it
20.3 – to advise the project manager on the practical implications of the design of the works and on subcontracting arrangements
20.4 – to prepare forecasts of the total actual cost for the whole of the works
26.4 – to submit the proposed contract data for each subcontract for acceptance
36.5 – to submit a subcontractor's proposal to accelerate for acceptance
52.2 – to keep records of costs and payments
52.3 – to allow the project manager to inspect accounts and records.

## Secondary option clauses

### Option G – Performance bond

G1.1 – to give the employer a performance bond for the amount stated in the contract data and in the form set out in the works information.

### Option H – Parent company guarantee

H1.1 – to give the employer a parent company guarantee in a form set out in the works information.

### Option J – Advanced payment to the contractor

J1.3 – to repay advanced payments to the employer in instalments as stated in the contract data.

## 7.13 Express obligations of the contractor

*Option R – Delay damages*

R1.1 – to pay delay damages as stated in the contract data from the completion date until completion or take over.

*Option S – Low performance damages*

S1.1 – to pay low performance damages as stated in the contract data for defects included in the defects certificate showing low performance.

*Option V – Trust fund*

V2.3 – to inform suppliers of the terms of the trust deed and the appointment of the trustees
 – to arrange that subcontractors ensure that their suppliers and subcontractors are similarly informed.

## 7.14 Express prohibitions on the contractor

**Core clauses**

21.2 – not to proceed with work until the project manager has accepted the design
26.2 – not to appoint a subcontractor without the project manager's acceptance
26.3 – not to appoint a subcontractor without the project manager's acceptance of the terms of the subcontract
30.1 – not to start work on site before the first possession date.

# Chapter 8

# Time (and related matters)

## 8.1 Introduction

Section 3 of the NEC is entitled 'Time', a title which hardly does justice to the variety of important matters contained in the section. These include:

- commencement
- completion
- programmes
- possession of the site
- access to the site
- facilities and services
- suspension of work
- take over
- use before take over
- acceleration.

Whether the simplicity and brevity with which all of these matters (except programmes) are dealt with are seen as a tribute to those who draughted them or as over ambitious pruning of traditional provisions may depend on the viewpoint of the observer. Like so much of the NEC, Section 3 may prove satisfactory so long as there are no problems, but deficient when there are.

**Commencement and progress**

Particularly conspicuous by their absence are any express requirements in the NEC that:

- the contractor should start on or about a particular date
- the contractor should proceed with due expedition and/or regularly and diligently
- the contractor should use his best endeavours to prevent or reduce delay.

## 8.1 Introduction

It may be the intention of the NEC that compliance with such requirements as these follows naturally from compliance with the detailed requirements on programming. But, detailed as the programme requirements are, they do not expressly require the contractor to match his progress with his programme. Consequently it is unlikely that the programme requirements of the NEC can be taken as indicators of binding contractual obligations on commencement and rates of progress.

Alternatively the NEC may be drafted in the belief that terms on prompt commencement and regular progress can be implied as a matter of general law and need not be expressed. But if that is the belief it is dangerously optimistic. The Court of Appeal decision in the case of *GLC v. Cleveland Bridge & Engineering Ltd* (1986) suggests that where there are no express terms on progress they will not be implied and that where the contract simply states the obligation of the contractor to be to finish on time then he is entitled to proceed at his own pace in doing so.

With the NEC the absence of any references to prompt commencement or rates of progress in the termination clauses of Section 9 adds to the argument that terms should not be implied.

There is, of course, a third possibility: that the NEC is intentionally indifferent to the contractor's rate of progress. But that is hardly compatible with the emphasis put on the programme or the broad policy of the contract.

### Damages for delay

Section 3 of the NEC is totally silent on the contractor's liability for damages (liquidated or otherwise) for late completion. The contract relies on the inclusion of secondary option R for liquidated damages or the application of the common law for unliquidated damages.

### Sectional completions

Similarly Section 3 is silent on sectional completions. For these it relies on the inclusion of secondary option L.

Note, however, that Section 3 does deal with possession and take over of a 'part' of the works – although 'part' is neither a defined term nor an identified term of the NEC.

### Suspension of work

The NEC does not use the phrase 'suspension of work' nor does it state in any detail the provisions which normally apply to suspensions. It simply gives the project manager powers to instruct the contractor to stop and to restart any work (clause 34.1). The financial implications of such an instruction are dealt with as a compensation event (clause 60.1(4)). Prolonged stoppages (exceeding 13 weeks) are grounds for termination (clause 95.6).

The contractor is not given the right, now common in other construction contracts, to suspend work in the event of prolonged late payment.

Where the NEC is used for process or plant contracts it should be noted that there are no provisions expressly dealing with delayed delivery to site. So unless these are added as special conditions the parties will have to rely on the general provision in clause 34.1.

**Take over**

In traditional contracts take over usually marks the point when responsibility for care of the works passes from the contractor to the employer. And that usually occurs when the contractor achieves completion or when the employer begins to use the works or parts of the works.

The NEC broadly follows traditional practice in that take over follows completion or use by the employer, but take over in the NEC lacks the formality it is given in some other contracts.

Unlike completion, take over is not a defined term of the NEC and its meaning is left to be determined from clause 35. Essentially it is a factual state of affairs which the project manager is required to identify by issuing a certificate. There is no mention in the clause of formal take over tests and, if required, they would have to be detailed in the works information or in special clauses.

An obvious purpose of identifying take over in the NEC is that after take over, loss or damage to the works is an employer's risk (clause 80.1). Another is that take over activates compensation event 60.1(15). But examination of the detail of clause 35 shows that it perhaps has a more fundamental purpose in that it allows parts of the works (as opposed to identified sections) to be treated as contractual entities.

However, it should be noted that although the NEC provides for take over of parts of the works, unlike most other contracts it does not provide that defects liability periods commence at take over. For further comment on this see Chapter 9.

## 8.2 Starting and completion

### Clause 30.1 – starting and completion

Clause 30.1 contains two important and distinct provisions:

- the contractor is not permitted to start work on the site until the first possession date
- the work is to be completed on or before the completion date.

Note that the first of these is a prohibition and not an obligation and that it relates only to work on the site. Obviously it is not intended to deter a start being made to design work.

Possession dates are identified by the employer in part one of the contract data for parts of the site and presumably (although it is by no means certain) the reference in clause 30.1 to 'the first possession date' means the first possession date for each part and not merely the first date listed in the contract data.

The brevity of the second provision in clause 30.1 – that completion shall be achieved on or before the completion date – has been touched upon in section 8.1 above. The provision says nothing on when the work is to commence or whether the contractor is to proceed regularly and diligently. It imposes simply the obligation to complete on time or before. It will, however, apply to sections if secondary option L is included in the contract.

One important point which comes out of the provision is the contractor's entitlement to complete early if he is able to do so. However, the contractual effect of this can be diluted if the employer exercises his right to state in the contract data that he is not willing to take over the works before the completion date.

Note that the wording used in clause 30.1 in relation to what has to be done before completion is 'does the work' and not 'provides the works'. This is consistent with clause 11.2(13) which defines completion as when the contractor has done 'all the work' which the works information states he is to do by the completion date. But its impact is that clause 30.1 relates only to such work as is expressly required to be completed within the set time and not to the more general obligation of the contractor to provide the works.

### Clause 30.2 – deciding and certifying completion

Clause 30.2, like clause 30.1, contains two important and distinct provisions:

- the project manager decides the date of completion
- the project manager certifies completion within one week of completion.

The true meaning of the first of these provisions is of great contractual significance. On ordinary reading the clause suggests that the parties have agreed that the project manager should decide the date of completion and that is the end of the matter so that it is not open to dispute. However, if it can be said that the project manager's decision is an action within the meaning of clause 91.1 (adjudication) or is otherwise not intended to be final, then it may be open to dispute. But note that there is nothing in clause 92.1 (the adjudicator) which expressly empowers an adjudicator to open up and review the project manager's decisions.

A point which will probably be raised if the matter goes to the courts, as it surely will unless the contract is amended, is that unless the project manager's decision is truly final the reference in clause 30.2 to the decision is superfluous since the obligation to certify completion fulfils by itself all contractual needs.

But quite apart from this important legal question there are some practical aspects to clause 30.2 to be considered. The problem is that the contractor has no entitlement under the NEC to apply for the completion certificate, less still an obligation to do so. It is not even clear how the project manager's decision is to be given. Consequently if the contractor and the project manager are in disagreement on whether or not completion has been achieved, the contract offers little guidance on procedure. Clause 13 on communications does not appear to take effect. And until the project manager has made his decision he is not in breach of his obligation to certify completion.

There is no provision in the contract for deemed completion and one effect of the second provision of clause 30.2, that the project manager shall certify completion within one week of completion, could be to throw into doubt the validity of any certificate backdated by a longer period than one week.

Where there is proven delay in issuing a completion certificate, and backdating may be evidence of this, the contractor may have a remedy under compensation event 60.1(18) – breach of contract by the employer – on the basis that the employer is responsible for the project manager's defaults.

## 8.3 The programme

### Clause 31.1 – submission of the programme

The NEC relies on there being an accepted programme.

The contractor may have identified a programme in the contract data, in which case that will normally become the first 'Accepted Programme'. Clause 31.1 requires that if there is no such identified programme then the contractor is to submit a first programme for acceptance within the period stated in the contract data. This is a period entered in part one of the contract data by the employer.

The contract does not deal expressly with the position if the project manager finds the programme identified in the contract data unacceptable. It is by definition the accepted programme. Presumably this is a matter to be dealt with before the award of the contract. And, having regard to the contractual effects of the accepted programme, particularly in the obligations it imposes on the employer, assessment of an identified programme is clearly an important pre-contract function.

**Clause 31.2 – detail of the programme**

Clause states 31.2 the detail to be shown on each programme that the contractor submits for acceptance. The amount of detail required by clause 31.2 is by any standards comprehensive and it disposes of any notion that a simple programme in bar chart form is adequate for the NEC. Instead it indicates that the term 'programme' as used in the NEC is not a single document but a collection of documents which may include method statements, histograms, bar charts, network diagrams and the like.

The details required are:

- Dates
  - the starting date (identified in the contract data)
  - possession dates (identified in the contract data)
  - the completion date (the completion date identified in the contract data or as revised).

- Method statements
  - for each operation
  - identifying the equipment and resources the contractor plans to use.

- Planned completion
  - the date by which the contractor plans to do all the work which the works information requires to be completed by the completion date. Clearly this can be in advance of the completion date.

- Order and timing of
  - operations the contractor plans to do in order to provide the works. (Some of these may precede the start of work on site and some may follow completion)
  - the work of the employer and others either as stated in the works information or as later agreed with the contractor.

- Dates
  - when the contractor plans to complete work needed to allow the employer and others to do their work.

- Provisions (allowances) for
  - float

- ○ time risk allowances
- ○ health and safety
- ○ procedures in the contract.
- Required dates for providing the works in accordance with the programme for
  - ○ possession of a part of the site if later than its identified possession date
  - ○ acceptances (certainly by the project manager – possibly by others)
  - ○ the supply of plant, materials and other things to be provided by the employer.
- Other information
  - ○ which the works information requires to be shown.

The point is not stated in clause 31.2 but it follows from other clauses that if Option L for sectional completions is used, the programme requirements for dates apply to sections and the whole of the works.

Failure by the contractor to submit a first programme for acceptance (where none is identified in the contract data) entitles the employer to retain one quarter of amounts due as interim payment (clause 50.3).

**Clause 31.3 – acceptance of programmes**

By clause 31.3 the project manager has to respond to submission of a programme within two weeks. The project manager must either accept the programme or state his reasons for not accepting it.

There is nothing in clause 31.3 to indicate whether or not deemed acceptance is intended if the project manager fails to respond within the time allowed. But since late response is a compensation event under clause 60.1(6), a late notice of non-acceptance is probably valid, and deemed acceptance, if it occurs at all, probably only occurs when there is no response whatsoever from the project manager.

In the event of non-acceptance of any programme the contractor is obliged to resubmit, taking account of the project manager's reasons, within the period for reply (clause 13.4).

Reasons stated in clause 31.3 for not accepting the programme are:

- the plans it shows are not practicable. (But note that this applies only to the contractor's plans and not to the plans of the employer and others in so far as they may be in the programme)
- information required by the contract is not shown. (This could be either information required by clause 31.2 or by the works information)
- the contractor's plans are not realistically represented. (This could obviously be highly contentious)

## 8.3 The programme

- it does not comply with the works information. (This probably means that the contractor's plans do not comply with the works information rather than that the programme itself is not compliant).

The project manager could withhold acceptance for reasons other than the above – perhaps to take account of changed circumstances relating to the employer or others – but non-acceptance for a reason other than the stated reasons is a compensation event (clause 60.1(9).

The effects of non-acceptance, other than to resubmit, are related mainly to the assessment of compensation events. Clause 64.2 allows the project manager to make his own assessment. Clause 50.3 on reduced interim payments does not necessarily apply to all non-acceptances since it refers only to submission of a first programme and to the absence of information.

The contractor is not expressly prohibited by the contract from proceeding in accordance with a non-accepted programme – although in extreme circumstances the consequence could be that the termination procedures of clause 95 might be invoked.

A less extreme scenario is that the project manager, unwilling to let the contractor proceed to a non-accepted programme, could give an instruction under clause 34.1 for the contractor to stop work. That would itself be a compensation event under clause 60.1(4) but not one entitling the contractor to any extra payment or time if the project manager decides that the reason for the stoppage was due to a fault of the contractor (clause 61.4).

There may even be circumstances where it is appropriate to put a dispute on non-acceptance of a programme to an adjudicator under clause 90.1. Indeed, it may be inevitable that if an adjudicator is called in to resolve certain disputes on the assessment of compensation events, he is obliged to consider arguments on the proper programme for the assessment.

Such arguments on the financial relevance of programmes are commonplace in construction disputes under all standard forms. But there is with the NEC, the potential for argument on programmes on a more practical level and some aspects of this are a cause for concern.

The stated reasons for non-acceptance in clause 31.3 include the project manager taking a view that the contractor's plans are not practicable or realistic. But that really should be a matter for the contractor to decide and not the project manager. Note that even if the contractor changes his plans to obtain the project manager's acceptance of his programme, that does not change the contractor's responsibility (clause 14.1). It might appear from this that the project manager has power without responsibility but project managers should be aware that if by their interference in the contractor's plans they cause loss to the contractor or worse, damage to persons or property, then they may, at least, render the employer liable for the financial consequences and, at worst, render themselves personally liable.

One interpretation of clause 31.3 which would to some extent act as a brake on the project manager's control over the contractor's programme is to take the word 'plans' in a narrow sense so as to exclude 'provisions' as listed in clause 31.2. This would then exclude from the listed reasons for non-acceptance supposedly inadequate 'provisions' for float, time risk allowances, health and safety requirements and other procedures.

With regard to 'float' in the contractor's programme, note that by clause 63.3 any delay to the contractor's planned completion gives an entitlement to extension of time. So as far as the NEC is concerned the old argument 'who does float belong to?' is firmly settled. It belongs to the contractor.

**Clause 31.4 – activities in the programme**

Clause 31.4 applies only in Options A and C – the contracts with activity schedules. It requires the contractor to include in each programme submitted for acceptance the start and finish of each activity in the activity schedule. This is additional to the requirements of clause 31.3.

## 8.4 Revising the programme

The NEC, as mentioned above, avoids expressly requiring the contractor to comply with his programme but it ensures that the contractor's programme does not become a redundant document by requiring the contractor to submit revised programmes at regular intervals or as instructed.

**Clause 32.1 – revised programmes**

Clause 32.1 states that the contractor is to show on each revised programme:

- progress achieved and timing of remaining work
- effects of compensation events and early warning matters
- plans for dealing with delays and notified defects
- other proposed changes to the accepted programme.

This detail is apparently intended to be additional to the detail required by clause 31.2; note the first line of that clause, 'The Contractor shows on each programme which he submits for acceptance'. Consequently prop-

## 8.4 Revising the programme

erly revised programmes are likely to be extensive and comprehensive documents – or rather sets of documents.

The requirement for actual progress to be shown is routine, as also is the requirement for its effect on the timing of the remaining work to be shown too. But note that with the NEC, planned completion as shown on the accepted programme (which as the works progress will be a revised programme) is the base for time related claims in compensation events – so with the NEC that base is mobile rather than fixed at commencement.

The requirement to show the effects of implemented compensation events appears to have few contractual implications because, once implemented, compensation events are not reassessed in the light of subsequent events (clause 65.2). However, the requirement to show the effects of notified early warnings – which will frequently be compensation events in the course of assessment (or not agreed) – may have contractual effects because it is clearly in the contractor's financial interests to underplay events for which he is responsible and to overplay events for which the employer is financially responsible. And bearing in mind that the project manager has only two weeks to respond to the submission of any programme (clause 31.3), he may have insufficient time to fully assess the implications of accepting a revised programme. See in particular the problem discussed under clause 32.2 below on multiple revisions.

The requirement for the contractor to show in revised programmes his plans to deal with delays and to correct notified defects appears to be a requirement to show resources rather than dates.

The final requirement of clause 32.1, that the contractor shall show other changes which he proposes to make to the accepted programme, comes fairly close to suggesting that the contractor is not entitled to depart from the approved programme without the project manager's acceptance. But its intention may be no more than to prevent the contractor unilaterally altering the employer's liabilities.

### Clause 32.2 – submission of revised programmes

Clause 32.2 states the circumstances in which a revised programme shall, or may be, submitted. They are:

- when instructed by the project manager to do so
- when the contractor chooses to do so
- at the intervals stated in the contract data.

However, see also clause 62.2 requiring a revised programme in quotations for compensation events.

In the case of an instruction, clause 32.2 requires the contractor to submit his reply within the period for reply stated in the contract data.

Unlike the provision in model form MF/1 there is nothing in clause 32.2 itself to indicate that the contractor should be paid the cost incurred in submitting a revised programme at the project manager's instruction. But arguably a compensation event can be claimed under clause 60.1(1) on the basis that the instruction is a change in the works information. Or the cost may be valued as part of the cost of a compensation event in appropriate circumstances.

The entitlement of the contractor under clause 32.2 to submit a revised programme when he chooses to do so has potentially serious implications for the employer and the project manager in present times of computerised programming. Some contractors are already deploying on-site programmers and estimators and are revising their programmes on a very regular (if not daily) basis to take account of compensation events and other matters. The burden this throws on the project manager who is obliged to respond to each revision is immense. But failure to respond is not only a breach of contract but also gives advantages to the contractor in the assessment of compensation events.

Routine revisions to the programme at the intervals stated in the contract data are only likely to be made in a contract with few compensation events since clause 62.2 requires a revised programme to be submitted in quotations for compensation events.

## 8.5 Shortened programmes

In the last 30 years or so a great deal has been written and said on the subject of programmemanship and its close relative, claimsmanship. The questions have been to what extent is it possible, or contractually permissible, for a contractor to devise a programme which is to his advantage in claims for extra time and payments.

Most of the debate has focused on the practice adopted by many contractors, for perfectly sound commercial reasons, of submitting shortened programmes showing completion well within the time allowed in the contract. Contractors see such programmes as necessary if they are to be competitive in tendering; but employers, not without some cause, may see them as props for opportunist claims or as unwelcome redefinement of their own liabilities or obligations.

**The Glenlion case**

From the decision in the much discussed case of *Glenlion Construction Ltd v. The Guinness Trust* (1987) it is clear that a contractor cannot rely on a shortened programme in claims for breach of contract. It is not open to one party to unilaterally change the obligations of the other party after the contract has been made. But *Glenlion* is not directly applicable to claims made under provisions of the contract. For such claims, where shortened programmes are relied on, it is necessary to see what obligations the employer has undertaken in the contract in respect of assisting or permitting the contractor to work to his programme.

Some contracts carefully avoid linking the employer's obligations to the contractor's programme and fix obligations in line with the common law principle of prevention, such that the employer must not hinder the contractor from finishing on time. Other contracts, and the IChemE Red Book is a good example, do fix the employer's obligations to the contractor's programme.

**Shortened programmes in the NEC**

The approach of the NEC on the employer's obligations is to fix them either in the works information or, if not so fixed, by reference to the contractor's accepted programme. Clause 60.1 listing the compensation events refers to both. But, in so far as it is left to the contractor to decide his own arrangements and fix his own timescale, the NEC approach is likely to encourage rather than diminish the use of shortened programmes.

However, with the NEC it is not simply the matter of fixing the employer's obligations which is relevant to the impact of shortened programmes. Another important matter is that the assessment of compensation events is determined by reference to the accepted programme. Contractors will not be slow to identify what advantages can be gained from this and it is difficult to avoid the conclusion that programmemanship will continue to thrive under the NEC much as before – perhaps more so.

## 8.6 Possession of the site

Clause 33 of the NEC deals with three matters:

- possession of the site
- access to the site
- the provision of facilities and services.

Neither possession nor access are defined terms of the NEC and they are therefore to be taken as having their ordinary meaning in so far as that is consistent with the contract.

Similarly neither facilities nor services are defined terms. But in the NEC the meaning of these terms can be derived from the works information which should state what facilities and services are to be provided by each of the parties to the other.

### Meaning of possession

Possession of the site under the contract does not mean that the contractor is literally the party in possession. The word 'possession' is used for convenience and it is not intended to confer on the contractor the full range of legal rights and liabilities which accompany possession by ownership.

Possession given to the contractor is more in the nature of a licence to occupy the site. Thus in *Nevill (HW) (Sunblest) Ltd* v. *William Press & Son Ltd* (1981) it was said:

> 'Although [the contract] uses the word 'possession' what it really conferred on William Press was the licence to occupy the site up to the date of completion'.

And in *Surrey Heath Borough Council* v. *Lovell Construction* (1988) the following passage from Hudson on Building Contracts (10th edition) was accepted as an accurate statement of the law:

> 'In the absence of express provisions to the contrary, the contractor in ordinary building contracts for the execution of the works upon the land of another, has merely a licence to enter upon the land to carry out the work. Notwithstanding that contractually he may be entitled to a considerable degree of exclusive possession of the site for the purpose of carrying out the works, such a licence may be revoked by his employer at any time, and thereafter the contractor's right to re-enter upon the site of the works would be lost. The revocation, however, if not legally justified, will render the employer liable to the contractor for damages for breach of contract, but subject to this the contractor has no legally enforceable right to remain in possession of the site against the wishes of the employer'.

### Meaning of access

The Concise Oxford Dictionary defines access as the 'right or means of approaching or reaching'. This clearly indicates, as of course is the case, that access can mean either the legal permission to enter or the physical needs for entry.

Contracts use the term 'access' in different ways, with some concerned only with the legal aspects and others concerning themselves also with the physical aspects. The IChemE Red Book goes so far as to require the employer to provide access from the nearest road or railway suitable for use by the contractor.

There is no fixed rule for distinguishing between the two types of access but the phrase 'to give access' (as used in the NEC) suggests the legal meaning, whereas the phrase 'to provide access' suggests a physical meaning.

**Clause 33.1 – possession**

Clause 33.1 requires the employer to give possession of 'each part of the Site' to the contractor on or before the later of:

- its possession date (which is a date identified in the contract data), *or*
- the date for possession shown on the accepted programme.

The reference in the clause to 'each part of the Site' is compatible with the layout of the contract data sheet but where the site is not divided into parts it seems to require, somewhat artificially, that the whole of the site is treated as a part of the site.

The obligation to give possession 'on or before the later of' the date in the contract data or the date in the contractor's programme, has two effects. First, it prevents the contractor from bringing forward in his programme the date of possession from the identified date. And secondly, it permits the employer to defer giving possession until the programme date.

Note however, there is nothing in clause 32.1 (revising the programme) to prevent the contractor from bringing forward a programmed date for possession in a revised programme. So it may be risky for the employer to rely on a programme date as a firm date for fixing his obligations.

In the event of acceleration under clause 36.1 there may be circumstances when it is necessary to bring possession forward of its identified date. The contract does not deal expressly with this situation but it can be accommodated by a separate agreement between the parties.

Failure by the employer to give possession as required by clause 33.1 is a compensation event (clause 60.1(2)). It is also a breach of contract and in extreme circumstances the contractor could be entitled to terminate the contract at common law.

Taken by itself there is nothing in clause 33.1 to suggest that the contractor is not entitled to exclusive possession of the site. But clause 25.1 also has to be considered. This requires the contractor to share the working areas with others as stated in the works information. And since the working areas include the site, it follows that the contractor does not have a right to exclusive possession of the site if the works information indicates otherwise.

**Clause 33.2 – access, facilities and services**

Clause 33.2 contains a mixed bundle of provisions which come into effect when the contractor has possession of 'a part of the site':

- the employer gives the contractor access to and use of that part of the site
- the employer and the contractor provide facilities and services as stated in the works information
- the project manager assesses any cost incurred by the employer as a result of the contractor not providing the required facilities and services
- the contractor pays the assessed cost (presumably to the employer).

Again it must be assumed that 'a part of the site' can be read as 'the whole or part of'.

The phrase 'gives — access to and use of' suggests an entitlement rather than something physical, particularly as facilities and services which are obviously physical are 'provided'. However, it is questionable whether such diverse matters should have been placed together in the same sentence.

Also questionable is whether it is wise that the NEC extends the usual obligation of giving 'access' to the site to giving 'use of'. It is not clear what is achieved by this; the contractor has separately been given possession and access, which should be sufficient. The implications of the contractor being permitted to use the site are considerable. Presumably use is intended to be related to the construction of the works, but even that could go beyond the legally permitted use of the site.

Surprisingly, having regard to the provisions for other defaults, there is no specific remedy in the contract for failure by the employer to give access to and use of the site. The only compensation event which appears to be relevant is clause 60.1(18) – breach of contract by the employer.

Failure by the contractor to provide facilities and services is expressly covered in clause 33.2; failure by the employer to provide something is covered by either compensation event 60.1(3) or 60.1(18). However, there is some doubt about 60.1(3) which may relate only to late provision.

Note the potential inequality in the respective contractual remedies available to the employer and to the contractor for non-provision of facilities and services. The employer recovers 'any cost incurred', the contractor only such amount as becomes due under the compensation event procedure – which may be considerably less than the actual cost incurred, particularly if some elements of the cost are of indirect nature.

## 8.7 Instructions to stop or not to start work

The point is made earlier in this book that the NEC is prone to condensing important and usually comprehensively drafted provisions into a single short sentence. The manner in which the NEC deals with the powers of the project manager to order suspensions of work in clause 34.1 is characteristic.

## Clause 34.1 – instructions to stop or restart work

Clause 34.1 states simply that the project manager may instruct the contractor to stop or not to start any work and may later instruct him to restart or start it.

The clause places no limitations on the project manager's powers, and the contractor's remedies for complying with the project manager's instructions are to be found elsewhere in the contract. The clause requires no explanation of the reasons for any suspension to be given.

If the procedures of the contract are properly followed it may be that some limitation of the project manager's powers is provided by clause 16.1 (early warning). In most cases it will be likely that an instruction to stop or not to start work will delay completion, so the project manager would seem to be under an obligation to call an early warning meeting before giving such an instruction. He would then be obliged to co-operate in making and considering proposals to avoid or reduce the effects.

Another limitation is, of course, regard for the employer's financial interests. And any instruction by the project manager to stop or not to start work is a compensation event (clause 60.1(4)); although, if the project manager decides that the event arises from a fault of the contractor and it is the contractor who has given notice of the compensation event, there is no change in the prices or the completion date.

The intention of the contract is probably that the contractor should be compensated for any instruction to stop or not to start work unless it can be proved that default by the contractor is the reason for the instruction. However, it is not certain that the wording of the contract actually achieves this.

One aspect of clause 34.1 which should not be overlooked is that by clause 95.6 any instruction to stop or not to start substantial work or all work, which remains in effect for 13 weeks, activates rights of termination. The employer may terminate if the instruction is due to a default of the contractor; the contractor may terminate if the instruction is due to a default of the employer; either party may terminate if the instruction is due to any other reason.

The problem with clause 95.6 is knowing what is meant by a default of the employer. Arguably it means only the contractual defaults in clause 95.1 (insolvency etc.) and in clause 95.4 (failure to pay on a certificate). If this is the case then employer's change of mind is not a default and termination would be resolved under the provisions for 'any other reason'. But see the comment in Chapter 15, section 2 on how this financially disadvantages the contractor.

Another aspect of clause 34.1 which requires consideration is whether it can legitimately be used by the project manager as a means of ordering omission variations. Strictly, an omission variation should require a change in the works information but note that the wording of clause 34.1

is discretionary and the project manager 'may' give an instruction to restart – he is not obliged to do so. This suggests that the clause can be used to reduce the scope of the work although it is doubtful if this is the intention of the contract.

## 8.8 Take over

General aspects of take over under the NEC are discussed in section 8.1 above. The detail of clause 35 is examined below.

A further general point to note on clause 35 is that it uses the term 'part of the Works' throughout. The NEC does not define the term 'part' but from the context of clause 35 it is clear that an identified section of the works can be regarded as a part for the purposes of the clause, and that the whole of the works is not excluded from the application of the clause but is to be regarded as the sum of the parts.

**Clause 35.1 – take over and possession**

Clause 35.1 states only the relationship between take over and possession of the site. It provides:

- that possession of each part of the site returns to the employer when he takes over the part of the works which occupies it
- that possession of the whole of the site returns to the employer when the project manager certifies termination.

The reason for the second provision is clear enough. On termination the employer effectively revokes the contractor's licence to occupy the site. The reason for the first provision is less clear and taken literally its effects are quite startling.

Normally the contractor retains rights of possession of the site until completion – an entitlement so obviously essential that most contracts do not find it necessary to state it. But under clause 35 of the NEC possession of the site is linked not to completion but to take over – and clause 35.3 expressly provides that take over precedes completion when the employer begins to use the works or any parts of them. This leads to the extraordinary possibility that the contractor can lose his right to occupy the site and complete the works even though he may not be in default in any way and even though the completion date may not have been reached.

It may be the intention of the NEC that use of the works (or parts) by the employer has conclusive effects in terminating the contractor's rights and obligations in respect of completion, but if that is the case (which seems unlikely) it would have been better if the point had been made more prominently and incorporated in the definition of completion in clause 11.2(13).

It is worth noting here that clause 43.3 does expressly require the employer to give access to, and use of, any part of the works taken over, to the contractor for correcting defects. But it is difficult to see on the wording how this has any application to work which is not completed when it is put into use or taken over.

### Clause 35.2 – take over and completion

A partial link between take over and completion is given in clause 35.2. But the clause deals (or is intended to deal) only with take over after completion and not the situation discussed above of take over before completion. Clause 35.2 has two distinct provisions.

The first sentence provides that the employer need not take over the works before the completion date if it is stated in the contract data that he is unwilling to do so. Clearly this is a decision to be made by the employer before the award of the contract. It covers the not uncommon situation where an employer perceives no benefit in early completion and does not want to take on, any earlier than planned, care of the works responsibilities.

The contractor remains entitled to finish early and remains entitled to have completion certified if he does so, but his responsibilities for care of the works continue until the set completion date.

It is unlikely that the provision in clause 35.2 is intended to take effect other than in the event of early completion. But since neither the provision itself nor the contract data entry mentions completion (both refer to the completion date), it is arguable that it could apply to use before completion, thereby negating the provision in clause 35.3 that the employer is required to take over the works when he begins to use them.

The second provision in clause 35.2 requires the employer to take over the works not more than two weeks after completion unless the employer has stated in the contract data that he is not willing to take over before the completion date. This supports the contractor's right to finish early and it gives the employer two weeks to arrange his own insurance cover.

### Clause 35.3 – take over and use of the works

Clause 35.3 permits the employer to use any part of the works before completion 'has been certified'. It then stipulates that such use is accompanied by take over except when:

- the use is for a reason stated in the works information, *or*
- the use is to suit the contractor's method of working.

It is not clear if any significance is intended by the use of the phrase 'Completion has been certified' in clause 35.3. The word 'Completion' would suffice unless the clause is taken to have a very restricted meaning such that the employer may use the works in between completion and the date when completion is certified.

Note that if the employer has stated reasons for not taking over parts of the works when they are put into use, he does so in the works information and not in the contract data. There appear to be no limitations on such reasons but they are probably intended to be of a practical nature.

But whatever the stated reasons they are at least settled at the time the contract is made and are hopefully dispute free. The same cannot be said for the second class of usage exempting the employer from take over. This is the class described as use 'to suit the Contractor's method of working'.

The difficulty here is in deciding whether a distinction should be made between a method of working which includes use which is unavoidable (e.g. in a road improvement scheme or the up-grading of a water treatment works) and a method of working which is solely of the contractor's choosing. It is not unreasonable, having regard to the contractor's responsibilities for care of the works and his liabilities for defects, that only the latter method of working should apply. But that is not always taken to be the case with similar provisions in other contracts and it is unlikely that everyone will agree that is the case under the NEC.

**Clause 35.4 – certifying take over**

Clause 35.4 requires the project manager to certify the date on which the employer takes over any part of the works within one week of the date of take over. The certificate is to show the extent of the part of the works taken over.

Where the project manager has certified completion this will not present any problems since by clause 35.2 the date of take over should not be more than two weeks after the date of completion.

Where completion has not been certified the operation of the clause will frequently depend on the project manager's decisions on whether the works have been put into use and whether the exceptions in clause 35.3 apply.

There is no specific remedy in the contract if the project manager fails to certify in accordance with clause 35.4, although this is, perhaps, another case where compensation event 60.1(18) – employer's breach of contract – might come into play.

It is, of course, arguable that when there is completion there is in any event 'deemed' take over under clause 35.2. This could suggest that clause 35.4 applies only to take over before completion but that is probably not what is intended.

## 8.9 Acceleration

The term 'acceleration' has various meanings according to the contract in which it is used and the circumstances in which it is used. Sometimes it means no more than the mitigation of delay; sometimes it means a formal shortening of the time allowed for completion. The NEC uses the term in the latter sense.

### Clause 36.1 – acceleration

Clause 36.1 states firstly, that the project manager may instruct the contractor to submit a quotation for acceleration to achieve completion before the completion date; and secondly, that a quotation comprises:

- proposed changes to the prices and the completion date, *and*
- a revised programme.

The clause is not to be taken as giving the project manager power to instruct acceleration. It is power only to seek a quotation. Acceleration, as described in the clause, invokes a change in the obligations of the parties which can only be made by agreement.

What is intended is that the contractor can be asked to state his price for taking on the contractual obligation of finishing early and putting in hand the necessary resources.

### Clause 36.2 – quotation for acceleration

The clause puts the contractor under an obligation to submit a quotation when so instructed or to give his reasons for not doing so. Reasons are to be given within the time for reply as stated in the contract data, but it is not wholly clear whether or not the same timescale is intended to apply to the submission of the quotation.

There is no sanction in the contract for failure by the contractor either to provide a quotation or to give his reasons. Strictly the contractor would be in breach of contract but it is unlikely that the employer would succeed in a claim for damages.

One reason that the contractor might offer for not submitting a quotation is that the contract does not provide for reimbursement of his costs in assembling the quotation. Some other standard forms such as MF/1 do provide for reimbursement.

When the contractor does submit a quotation there may be some question as to whether he is bound by the compilation rules applying to quotations for compensation events. The terminology in clause 36.1 matches that in clause 62.2 and no doubt it would assist the project manager in assessing the quotation for standardisation to apply. However, the imposition of a link between acceleration quotations and compensation event quotations cannot be enforced even if it is intended.

**Clause 36.3 – accepting a quotation**

Clause 36.3 is found in main options A, B, C and D. It states that when the project manager accepts a quotation for acceleration, the completion date and the prices are changed and the revised programme is accepted.

In short, acceptance of the quotation concludes the formalities of changing the contractual obligations. Clearly it is assumed that the project manager is acting with the full authority of the employer.

**Clause 36.4 – accepting a quotation**

Clause 36.4 applies to acceptance of quotations for acceleration under main options E and F. It states that when the project manager accepts a quotation he changes the completion date and accepts the revised programme.

The difference from clause 36.3 applying to main options A, B, C and D is that the reference to changing the prices is omitted. This is because Options E and F are fully cost reimbursable and the prices are defined in terms of cost.

**Clause 36.5 – subcontractors**

Clause 36.5 applies only to main options C, D, E and F – those options when a subcontractor's costs are defined as part of actual cost. The clause states that the contractor submits a subcontractor's proposal to accelerate to the project manager for acceptance.

In other words, when the contractor is being reimbursed on a cost basis, and the contractor's quotation for acceleration includes for subcontractor costs, those costs must meet with the approval of the project manager.

# Chapter 9

# Testing and defects

## 9.1 Introduction

The NEC deals very briefly with testing and defects. In clauses 40 to 45 it sets out, in just one and a half pages, the basic obligations of the parties and the role of the supervisor. But the NEC does not attempt to match the detailed provisions on quality, quality control, testing and defects found in standard construction, process and plant contracts.

This is a matter of policy and not one of neglect. In the interests of flexibility the detail for individual contracts is left to be supplied in the works information. Perhaps in recognition of the potential dangers of this approach, no fewer than six pages of the Guidance Notes and eight pages of the flow charts are devoted to testing and defects.

The principal danger for users of the NEC, particularly new users, is that they may not recognise that it is insufficient to specify types of tests and acceptable results in the works information. That may be adequate for many conventional contracts but it will not do for the NEC.

Because the NEC relies so heavily on what is 'required by the Works Information', for testing it is, for example, necessary to specify in the works information:

- what tests apply
- who undertakes tests
- when tests are to be done
- where tests are to be done
- who provides materials, facilities, samples etc.
- what procedures apply
- what standards apply.

There is, of course, a secondary danger in this in that there will be a temptation to include in the works information extracts from familiar standard forms written in a different style and with a different philosophy from the NEC. But it may be even more dangerous for individuals to attempt to copy the drafting style of the NEC. On balance it is probably best to write the works information in straightforward language which is meaningful to the technical people who are going to be concerned with testing and defects.

Not everyone is enthusiastic about the way the NEC deals with testing and defects. One objection which is put quite strongly is that the expertise of users of standard forms is developed through experience of consistent procedures and these are lacking in the NEC. Another objection is that the costs of preparation of the contract documentation are increased because of the amount of detail to be drafted into the works information.

Whatever the weight of those dangers and objections in the construction industry, there is little doubt that they are matters of serious concern in the process and plant industries. Users of the IChemE forms and the IMechE/IEE forms place great reliance on the detailed provisions in their forms for tests before delivery, tests on completion (take over tests) and performance tests. Their initial response to the approach in the NEC has been less than favourable.

**Types of tests**

Unlike other standard forms of contract the NEC does not distinguish between different categories of test. Depending on what is required in the works information for any particular contract, 'tests' within the meaning of the NEC may be:

- tests before delivery
- tests before completion
- tests on completion (or take over)
- performance tests.

The standard provisions in clauses 40 to 45 of the NEC apply to all specified tests.

This calls for great care in the drafting of the works information, particularly in regard to role reversal for performance testing. Most contract testing is carried out by the contractor, with the employer (or his representative) supervising or observing. But as a general rule performance testing is undertaken by the employer because it follows take over and the employer is in possession of, in control of, and is using the works. The role of the contractor is then to supervise or observe.

The NEC does not expressly deal with role reversal because it does not mention performance tests in its core clauses. In fact the only specific mention in the NEC to performance is in Option S (for low performance damages). This does no more than state the contractor's obligation to pay any low performance damages specified in the contract if a relevant defect is included in the defects certificate.

**Role of the supervisor**

The principal role of the supervisor under the NEC is to monitor quality and performance on behalf of the employer. Consequently most of the

## 9.1 Introduction

references to the supervisor are found in the clauses on testing and defects.

In clauses 40 to 43 it is the supervisor and not the project manager who exercises control over the contractor. It is the supervisor who undertakes the important tasks of notifying defects and issuing the defects certificate. However, in clauses 44 and 45 relating to accepting defects and uncorrected defects, control reverts to the project manager.

To the extent that the supervisor and the project manager act in independent professional capacities, there is some potential for conflict in this. On the face of it the project manager alone decides whether or not to accept defects. However, there must be a presumption that the supervisor is at least consulted.

### Obligations and entitlements

A practical question which frequently arises on construction, process and plant contracts is whether, in the event of the contractor failing in his obligations to undertake tests or remedy defects, the employer is entitled to carry out such work. The contractual answer in most standard forms is yes, subject to notice being given by the employer, and the contractor being given time to remedy the situation. And usually that applies both before and after completion.

The NEC does not directly address the issue except in relation to the remedying of defects after completion. It is arguable therefore that under the NEC, with the absence of any express entitlement for the employer to act prior to completion, to do so would be at the employer's risk and expense and possibly a breach of contract.

The sanction on the contractor for his default could in extreme cases rest in the termination provisions of the contract. But at a more commonplace level sanctions would operate by affecting the amounts in the contractor's entitlements to interim payments.

### Defects under the NEC

The NEC uses the word 'defect' with a strictly limited and defined meaning. Essentially it refers only to defects which are in some way the fault of the contractor and for which the contractor is contractually responsible. The contract is generally silent on defects for which the employer is responsible and the intention is probably that the project manager should give instructions dealing with such defects.

But there is, perhaps, an assumption in this approach to defects that responsibility can readily be allocated or apportioned. In reality, of course, the opposite is frequently the case. A taxing question for the NEC will be whether its rigidly structured provisions can accommodate the uncertainties and arguments which are a normal feature of the examination of causation of defects and the making of decisions on their treatment.

**Latent defects**

The NEC does not expressly exclude the contractor's liability for defects which only appear after the defects certificate has been issued – defects which may properly be called latent defects. Nor for that matter do the standard forms of construction contract. The general position is that if defects appear for which the contractor is responsible, the contractor is in breach of contract and an action for breach can be commenced within the appropriate limitation period. Under English law that is 6 years for ordinary contracts and 12 years for contracts executed as a deed (previously called contracts under seal).

However, the position is somewhat different under standard process and plant contracts and most expressly exclude the contractor's liability for latent defects.

Under the NEC, because there is no exclusion of liability for latent defects (and there is no certificate issued which is stated to be conclusive evidence of fulfilment of the contractor's obligations), the contractor's liability for latent defects follows the law applicable to the contract (subject to any limitation arising from clause 21.5). This is a point to be considered when the NEC is used for process or plant contracts.

The limitation in clause 21.5 applies only to defects due to the contractor's design – not workmanship – and it operates only when the amount of any limit is stated in the contract data.

**Option clauses**

The testing and defects provisions in core clauses 40 to 45 of the NEC apply unchanged in all of the main options A to F.

No secondary option clauses refer to testing and the only secondary option clauses which make reference to defects are:

- Option M – clause M1.1 – limitation of the contractor's liability for design
- Option P – clause P1.2 – retention
- Option S – clause S1.1 – low performance damages.

## 9.2 Definitions and certificates

### Definition of 'Defect'

Clause 11.2(15) defines the meaning of 'Defect' within the NEC. A 'Defect' is:

- a part of the works which is not in accordance with the works information, *or*
- a part of the works designed by the contractor which is not in accordance with:
  - the applicable law, *or*
  - the contractor's design which has been accepted by the project manager.

This is an interesting definition in two respects. Firstly, it excludes defects due to design for which the employer is responsible. Secondly, it appears to relieve the contractor of responsibility for defects in his own design once that design has been accepted by the project manager – subject to the design being in accordance with the works information.

### Defects which are not a 'Defect'

The first point above raises the question of whether the contractor has any obligation to remedy defects which are due to the employer's design. There is nothing in clauses 40 to 45 to cover this but one possible answer is that if it can be said that the defect constitutes damage to the works, then the contractor has an obligation to repair under clause 82.1 (repairs).

Another possible answer is that defects which are due to the employer's design are intended to be dealt with by way of project manager's instructions. The contractor is then obliged to comply and make good and the compensation event procedure comes into force.

Just how well this would operate in the event of a dispute on whether or not a defect is or is not a 'Defect' within the contractual definition, is not clear. Such a dispute, which is likely to be commonplace, would raise questions on the authority of the supervisor to notify the defect and the obligations of the contractor to take any remedial action until in receipt of an instruction from the project manager. However, the statement in clause 90.2 (settlement of disputes) that the parties must proceed as if a disputed action is not disputed until the matter in question is settled by adjudication, could be relevant. If it is relevant the contractor would be obliged to remedy the disputed defect on the notification of the supervisor and then seek recompense through adjudication.

## Acceptance of the contractor's design

As to the implications of the reference in clause 11.2(15) to 'the Contractor's design which has been accepted by the Project Manager', this is likely to be a long-running source of contention. The question, mentioned above, is to what extent the contractor is relieved of responsibility for defects in his design by the project manager's approval.

It may be possible to dismiss the suggestion of any such relief by reference to clause 14.1 – the project manager's acceptance does not change the contractor's liability for his design. The other argument for dismissal is that design which is defective will not be in accordance with the works information and will therefore be a defined 'Defect' under the first limb of clause 11.2(15). But there will be many occasions when this will not hold good, for example, when the works information is written in general terms or as a performance specification.

From the employer's viewpoint it is difficult to see what useful purpose is served by the words 'which has been accepted by the Project Manager' in clause 11.2(15). Employers who are in doubt on the matter consult their lawyers about possible deletion.

## Definition of defects certificate

Clause 11.2(16) defines what is meant by a defects certificate within the NEC. A defects certificate is either:

- a list of defects that the supervisor has notified before the defects date which the contractor has not corrected, *or, if there are no such defects*
- a statement that there are none.

At first sight it might appear that the defects certificate is the equivalent of the defects correction certificate under the ICE 6th edition conditions of contract or the certificate of completion of making good defects under JCT 1980. But it is not. Those certificates are only issued when the contractor has fulfilled his obligations to make good defects, whereas under the NEC the defects certificate is issued on a set date as a record of whether or not the contractor has fulfilled his obligations.

Note, however, that due to the restricted definition of 'Defect', the defects certificate does not cover defects due to the employer's design. Nor, due to its own definition, does the defects certificate cover defects noticed and notified by the supervisor after the defects date.

The definition of defects certificate in its reference to 'a statement that there are none' is not precise as to who should issue this statement or what form it should take. Presumably the intention is that the supervisor should issue a formal document headed defects certificate which makes

## 9.2 Definitions and certificates

the statement. This would seem to follow from clause 43.2 which requires the supervisor to issue the defects certificate. However, it is interesting to contemplate what the contractual position might be if, in the event of failure or delay by the supervisor to issue a defects certificate on the required date (a common situation under most standard forms of contract), the contractor himself issues a formal statement that there are no defects.

### Effects of the defects certificate

The defects certificate is mentioned in the following clauses of the NEC:

- clause 11.2(16) – definition
- clause 21.5 – limitation of the contractor's design liability
- clause 43.2 – issue of the defects certificate
- clause 50.1 – assessment for payment
- clause 80.1 – employer's risks
- clause 81.1 – contractor's risks
- clause 82.2 – repairs
- clause 84.2 – insurance cover
- clause P1.2 – retention
- clause S1.1 – low performance damages
- clause V3.1 – trust deed.

The purpose of the defects certificate as mentioned above is to put on record the state of the works at the date at which the contractor's entitlement to make good defects expires. It is not therefore to be taken as a certificate of confirmation of fulfilment of the contractor's obligations.

However, the effect of the defects certificate is similar to that in other standard forms in that it triggers the release of the final tranche of retention money and sets the date for expiry of various obligations.

A point to note is that it is the date of issue of the defects certificate which is generally effective in the above listed clauses and the implication from clause 43.2 is that the certificate will be issued by the supervisor immediately it becomes due.

### Defects date

The defects date is not a defined term of the NEC. It is not even a fixed date. It is a date to be determined from a period entered by the employer in part one of the contract data. That period indicates how many weeks after completion of the whole of the works the defects date occurs. Thus, instead of stating in the usual manner that there is a defects liability period of say six months or 12 months, the NEC arrives at a similar position by reference to its defects date.

If, by operation of the compensation event procedure, the time for completion of the works is extended, then the defects date is correspondingly adjusted.

The defects date has three principal express purposes in the NEC:

- under clause 42.2 it sets the date until which the contractor and the supervisor are obliged to notify each other of 'defects' which they find
- under clause 43.2 it sets (with some adjustment for the last defect correction period) the date for issue of the defects certificate
- under clause 61.7 it sets the final date for notification of a compensation event.

But what the defects date also does, although not expressly stated, is set the period within which the contractor is entitled to access to the site to remedy his own defects (again with some adjustment for the defect correction period). This can be implied from the obligation to remedy defects stated in clause 43.1.

**Defect correction period**

The phrase 'defect correction period' as used in the NEC has a wholly different meaning from the identical phrase used in the ICE 6th edition conditions of contract and such phrases as 'defects liability period' and 'maintenance periods' used in other standard forms.

In the NEC the defect correction period is a period of weeks entered in part one of the contract data by the employer to indicate how long the contractor is given to remedy 'defects' after completion. It is not the whole of the period from completion to the defects date. There is no definition in the NEC to this effect but again it can be deduced from the wording of clause 43.1. That clause states the contractor's obligation to correct notified 'defects' before the end of the defect correction period, and states that the period begins at completion for defects notified before completion.

## 9.3 Tests and inspections

### Clause 40.1 – tests and inspections

This brief clause does no more than define the scope of clause 40. It states that clause 40 only governs tests required by the works information 'and' the applicable law. It is probable that the word 'and' should read 'or'. Otherwise, taken literally, the clause is of very limited application.

But in any event, even taking the wider application, there is no doubt that clause 40 is intended to be restricted in its operation to specified or statutory tests. That is, it does not apply to tests and inspections which the contractor or the supervisor carry out for their own purposes. This emphasises the need for comprehensive testing and inspection requirements to be detailed in the works information.

The project manager does have the power to change the works information (clause 14.3) and therefore can issue instructions on additional or varied tests. They will then be treated as compensation events (clause 60.1(1)).

The supervisor, however, has no express power to change the works information and accordingly has no obvious power to order additional or varied tests.

### Clause 40.2 – materials, facilities and samples

Clause 40.2 is another brief clause. It simply confirms the obligations of the contractor and the employer to provide materials, facilities and samples for tests and inspections as stated in the works information.

The only point to note is that yet again these obligations take effect only to the extent that they are detailed in the works information. So unlike the position in most other standard forms, there are no fixed obligations on either party in respect of certain categories of tests and inspections.

### Clause 40.3 – notifications

Clause 40.3 deals with notifications of intentions to test and inspect and notification of results. It also confirms the power of the supervisor to watch any test by the contractor.

Both the contractor and the supervisor are required to notify each other before commencing tests or inspections and to notify each other of results afterwards. There could be some argument on whether 'results' means simply pass or fail or whether it means full records. There should be no argument, however, on whether the notifications need to be in writing. Clause 13.1 requires all notifications to be in writing.

With regard to tests or inspections which the supervisor is to carry out, clause 40.3 expressly requires the contractor to give notice before doing work which would obstruct the tests or inspections. Failure by the contractor would be a breach of contract but the employer's remedy is not immediately apparent. The supervisor could arguably give instructions to uncover under clause 42.1 (searching and notifying defects), but neither the project manager nor the supervisor has the express power to order the contractor to break out premature work (or even to break out patently faulty work).

There is, however, one firm sanction to encourage the contractor to give timely notice and that is the proviso in clause 60.1(10) – the compensation event for searching. The proviso makes it clear that the compensation event does not operate if the contractor gave insufficient notice of doing work obstructing a required test or inspection.

The final provision in clause 40.3, that the supervisor may watch any test done by the contractor, might appear to be superfluous as an obvious implied term but at least it puts the matter beyond doubt. Additionally it can perhaps be taken as giving the supervisor the right to attend off-site tests and the entitlement to the necessary facilities to observe on-site tests.

### Clause 40.4 – repeat tests and inspections

This is a short clause which in most contracts would be straightforward in its application and free from complication. It states that if a test or inspection shows that any work has a defect, the contractor is to correct the defect and repeat the test or inspection.

The problem with the NEC is that defect has a restricted meaning and tests and inspections are only those required by the works information (or statute). The clause is not therefore of general application to defective work. It is of application only to a category of defect revealed in a particular way.

In some circumstances where the clause does actually apply, there may be practical or financial reasons for the contractor to propose that the defect should not be corrected. Clause 44.1 (accepting defects) then comes into play. For comment on this see section 9.7 below. The contractor would also be obliged to give an early warning notice under clause 16.1 if the defect could delay completion or impair the performance of the works.

### Clause 40.5 – the supervisor's tests and inspections

Clause 40.5 deals with the possibility of the supervisor's tests and inspections causing delay to the works or payment to the contractor. The clause requires the supervisor to avoid unnecessary delay to either.

The obligation to avoid delay is reinforced by the compensation event in clause 60.1(11) which deals expressly with any test or inspection done by the supervisor causing unnecessary delay.

It is not clear how, within the context of the contract, an unnecessary delay is to be distinguished from a necessary delay. In the event of dispute the matter would have to be put to the adjudicator.

Nor is it fully clear whether the word 'delay' as used in the compensation event is meant to have its normal meaning of delay to the works or a wider meaning including delays to payments. The point is important because the provisions in clause 40.5 on delayed payments seem to suggest no more than that if the supervisor fails to carry out tests or inspections which control the contractor's entitlement to payment, then the contractor becomes eventually entitled to payment at the defects date (subject to adjustment for the last defect correction period) whether or not the supervisor carries out the tests or inspections – providing always that the delay is not the contractor's fault.

This, by itself, is a very poor remedy for the contractor who is being kept from his money, and it could have the potential for abuse in the wrong hands. It is even doubtful if the adjudicator could provide relief to the contractor for a payment delayed by the supervisor's lack of action, because the wording of clause 40.5 is precise in stating when (belatedly) the payment eventually becomes due.

### Clause 40.6 – costs of repeat tests and inspections

Clause 40.6 relates to costs incurred by the employer where tests or inspections have to be repeated after a defect is found. The clause states that the project manager assesses the cost incurred and the contractor pays the amount assessed.

Similar clauses are found in process and plant contracts but rarely in construction contracts. However, such clauses normally refer to the employer's costs in observing or supervising the contractor carrying out repeat tests. Clause 40.6 seems to apply to tests carried out by the employer, but this may not be the intention.

It is certainly unlikely that the intention of clause 40.6 is that it should be applied to any serial testing which might be thought necessary after the discovery of a defect. But, taking a broad view of the meaning of 'repeated', it could be argued that the clause allows for such testing.

## 9.4 Testing and inspection before delivery

### Clause 41.1 – delivery of plant and materials

Clause 41.1 is a prohibitive clause, of a type common in process and plant contracts, aimed at ensuring that all specified off-site tests and inspections are undertaken before delivery of plant or materials.

The clause states that the contractor shall not bring plant and materials to the 'Working Areas' until the supervisor has notified the contractor that they have passed tests or inspections required by the works information. 'Working Areas' is a defined term of the NEC and it includes both the site and any additional areas to be used by the contractor and detailed in part two of the contract data.

The clause is somewhat unusual in that instead of the contractor notifying the supervisor that tests and inspections have been passed, the position is reversed and it is for the supervisor to notify the contractor. It might be implied from this that the clause only applies to tests and inspections which the works information requires to be carried out by the supervisor. But that would seriously emasculate the clause since the majority of pre-delivery tests and inspections are carried out by specialist subcontractors and suppliers.

However, there is clearly a danger in the supervisor taking on the burden of notifying the contractor that tests and inspections have been passed (and accordingly plant and materials are fit for delivery) if the supervisor has not been responsible for those tests and inspections.

Obviously the clause can be applied without too much difficulty to tests and inspections carried out by independent agencies engaged by the employer, since the supervisor can then take on the role of examining results before they are passed to the contractor. However, for tests carried out by the contractor, his subcontractors or his suppliers, it is difficult to find any meaning in the clause other than that the contractor should supply the supervisor with results and should then await the supervisor's consent to commence delivery.

## 9.5 Searching and notifying defects

### Clause 42.1 – instructions to search

This clause gives the supervisor a broad power to instruct the contractor to search. However, the supervisor is required to give his reasons for the search with his instruction.

'Searching' is not a defined term in the NEC but clause 42.1 lists various actions which come within the scope of searching for the purposes of the clause. These include:

- uncovering, dismantling, re-covering and re-erecting work
- providing facilities, materials and samples for tests and inspections done by the supervisor
- doing tests and inspections which the works information does not require.

In most instances searching will be associated with an apparent or a suspected defect but the clause does not expressly limit itself to searches for defects. This may be because of the way the NEC avoids reference to 'defects' in the general sense and uses instead in its provisions the defined term 'defect' with its restricted meaning. Clearly a clause confined to searching narrowly for defined 'defects' rather than for defects in a more general sense would be inadequate.

But in the absence from clause 42.1 of reference to either class of defect it does open up the possibility that the supervisor is entitled to instruct the contractor to search for any reason whether or not connected with a defect. That then gives the actions listed as 'searching' in clause 42.1 potentially wide application. Thus uncovering work could be instructed for no other reason than the contractor's failure to give adequate notice of covering-up. Providing facilities etc. under the clause could include instructions for the contractor to supply routine testing equipment or to make other arrangements not covered by the works information.

And the final 'searching' action listed in clause 42.1 – 'doing tests and inspections which the works information does not require' – could be used as a general power by the supervisor to supplement the requirements in the works information for testing and inspections. This, of course, may be stretching the clause well beyond its original intentions and the probability is that all the actions listed as 'searching' are meant to apply only to searches for defects.

However, whatever the interpretation put on the clause, the supervisor and the contractor should both be mindful of the compensation event in clause 60.1(10). This applies when the supervisor instructs the contractor to search and no defect is found (except when the search is needed because the contractor gave insufficient notice of doing work obstructing a required test or inspection).

Consequently, although the supervisor may be able to use clause 42.1 as a general extension of his powers, the contractor will always be able to claim under the compensation event procedure for any instructions given unless a defect is discovered as a result of compliance with the instructions.

### Clause 42.2 – notification of defects

Clause 42.2 deals with the notification of defects. It requires both the supervisor and the contractor to notify each other of any defects they find.

The clause is unusual compared with provisions in other contracts in that it places a contractual obligation on the contractor to give notice of his own defects. But within the framework of the NEC this is understandable and is in keeping with the early warning procedure.

Note that it is only defined defects which have to be notified under clause 42.2 and not defects in the wider sense. When the contractor finds a defect he has to form a judgment as to whether it is a defect for which he is responsible (a Defect) or a defect due to employer's design or some other cause which places it outside the definition of defect. He then has to decide whether to give notice under both clause 16.1 (early warning) and clause 42.2 (notifying defects) or just under one clause. The danger for the contractor in giving notice under clause 42.2 without first considering the cause of the defect is that it could be taken as an admission of responsibility for the defect.

For the supervisor there is no obligation (or any power) to give early warning notice under clause 16.1 in the event of discovery of a defect due to employer's design. That obligation rests with the project manager. There must be an implied procedure therefore that the supervisor liaises with the project manager on early warnings.

One practical aspect of the notice requirements of clause 42.2 which may cause some concern is that, like other requirements in the NEC, they are expressed as always binding – without regard to the scale or importance of the matters involved. Taken literally, clause 42.2 could lead to an endless stream of written notices between the supervisor and the contractor for every minor infringement of the specification as the works proceeded.

Given the possibility of such a situation the contractor might well be inclined to argue that a defect cannot exist as such until the works are offered up as complete. Such an argument is not obviously at odds with the definition of defect in clause 11.2(15). Nor is it wholly at odds with the opening words of clause 42.2, which say 'until the defects date', if that is read as applying to a period from completion until the defects date. Moreover it is an argument which fits in well with the provisions in clause 43.1 for the correction of defects.

## 9.6 Correcting defects

### Clause 43.1 – correcting defects

Clause 43.1 serves two purposes. It states the contractor's obligations to correct defects and it defines when the defect correction period begins. Like clause 42.2, clause 43.1, refers only to defined defects and not to defects in general.

Firstly, clause 43.1 states that the contractor is to correct defects whether or not the supervisor notifies them. The purpose of the reference to the supervisor is not wholly clear. It may be an indicator that the clause applies only after completion. It would certainly appear to go without saying that the contractor is required to correct known defects for which he is responsible before completion. He would not be able to achieve completion without doing so. And there would be very little weight in any argument that the contractor was not obliged to correct defects known before completion unless notification was given by the supervisor.

## 9.6 Correcting defects

As to defects occurring or discovered after completion it is normal in most contracts for the contractor to be given notice, if only for the practical reason that he may not otherwise know about them. Clause 43.1 appears to suggest that the NEC takes a different approach, but when it comes to clause 45.1 (uncorrected defects) it can be seen that there is an express sanction only for failure to correct notified defects. Perhaps this provides another explanation for the reference to the supervisor in the opening sentence in clause 43.1, namely that not only the supervisor but also other persons such as the project manager and the employer may give notice of defects after completion.

The second sentence of clause 43.1 states that the contractor corrects notified defects before the end of the defect correction period. That period, as indicated in the third sentence of the clause, begins at completion for defects notified before completion; and for defects notified after completion it begins when they are notified. This means that the contractor is put under an obligation to correct notified defects after completion within whatever timescale is stated as the defect correction period in part one of the contract data. This is a sensible arrangement which will be welcomed by most employers.

For defects which are not notified, the position as indicated above is not clear. Clause 43.1 imposes an obligation on the contractor to correct them, but not within any particular timescale. However, there is no sanction in the contract for failure to correct such defects (clause 45.1 applies only to notified defects) and they cannot be included in the defects certificate.

As for the timing for the correction of defects before completion that is left to the contractor. But it is subject always to the discipline imposed by clause 11.2(13) which states that completion is when the contractor has done all the work required and has corrected notified defects which would have prevented the employer from using the works.

### Clause 43.2 – issue of the defects certificate

Clause 43.2 states briefly the obligation of the supervisor to issue a defects certificate. This must be done at the later of,

- the defects date, *or*
- the end of the last defect correction period.

Unlike some standard forms of contract, the supervisor is not given a period of grace within which to issue the certificate. The requirement is that it should be done on the due day.

Delay in issue of the certificate would leave the employer open to a claim for damages based on prolonged risk carrying by the contractor under clauses 81, 82 and 84 and late release of retention under clause P1.2.

However, as discussed in section 9.2 above, the NEC defects certificate is wholly different from a conventional defects correction certificate and there is no reason why there should be any of the usual delay (due to arguments on outstanding remedial works) in the issue of the certificate.

### Clause 43.3 – access for correcting defects

Clause 43.3 covers the contractor's entitlement to access to the works after they are taken over in order to correct defects.

The project manager is required to arrange access and the employer is required to give access. If the project manager fails to arrange access within the defect correction period he is obliged to extend the period for correcting the defect.

Note that, strictly, the clause does not say that the project manager extends the 'defect correction period'; it says that he extends the period for correcting the defect. It could be argued that these are two different things and there are indeed different contractual consequences depending upon which view is taken. One leads to a delay in the issue of the defects certificate, the other does not.

The questions then to consider are whether the contractor should be compensated for delayed access and possibly delay in issue of the defects certificate. There is no compensation event dealing with an authorised delay in the issue of the certificate and the contractor might have difficulty in arguing breach of contract having regard to the wording of clause 43.3.

## 9.7 Accepting defects

### Clause 44.1 – proposals to accept defects

Clause 44.1 is a very useful practical provision setting in train the possibility of acceptance of a defect if the contractor and the project manager so agree. The clause simply permits either to propose to the other that the works information should be changed so that the defect does not have to be corrected.

There is no mention in the clause of the supervisor. This is presumably because any agreement which follows from the clause is effectively an agreement between the parties (with the project manager acting for the employer) and because any agreement will require a change in the works information and only the project manager is empowered to make such a change (clause 14.3). However, it can be expected that the supervisor will have a behind the scenes involvement in the process.

There is nothing in clause 44.1 to indicate when it can be put into operation and it would seem to operate both before and after completion.

## Clause 44.2 – acceptance of defects

Clause 44.2 details the procedure for the acceptance of a defect.

The first step is that both the contractor and the project manager must indicate that they are prepared to consider the change. That can be gathered from the opening sentence of the clause. The question is, are either under any obligation to consider the change? The answer is far from simple.

It is too easy to say that the employer is entitled to strict performance to the works information and will consider nothing less. The common law will not necessarily support this. See, for example, the House of Lords decision in *Ruxley Electronics* v. *Forsyth* (1995). And in any event the contractual provisions when taken as a whole may indicate a different position.

In the case of the NEC the first relevant contractual provision is that found in clause 10.1 – the parties are obliged to act in a spirit of co-operation. Not being prepared to even consider a proposal from the other party would surely be a breach of this. The second relevant provision is in clause 45.1 (uncorrected defects) which implies that the contractor is liable for the full correction cost of any defect whether or not the correction is actually made by the employer. This is potentially so severe that again it leads to the view that the employer is obliged to consider proposals made under clause 44.1.

The second step under clause 44.2, assuming that the contractor and the project manager are prepared to consider proposals for change, is that the contractor submits a quotation for reduced prices or an earlier completion date or both. The third step is that if the project manager accepts the quotation, he gives an instruction to change the works information, the prices and the completion date accordingly. Note that under clause 60.1(1) such an instruction is expressly not a compensation event.

In the event of a dispute on the acceptability of the contractor's quotation that would appear to be a matter which could quite properly be referred to adjudication under clause 90.1.

## 9.8 Uncorrected defects

### Clause 45.1 – uncorrected defects

Clause 45.1 is, in broad terms, similar to the type of clause found in every construction, process and plant contract entitling the employer to recover the cost of making good uncorrected defects. On strict interpretation of its wording, however, it may be far more onerous on the contractor than any other standard form.

The clause says that if the contractor has not corrected a notified defect within its defect correction period the project manager assesses the cost of having the defect corrected by others and the contractor pays this amount.

The point to note is that the contractor is not simply liable for the cost incurred in correcting the defect, as is the normal position. Under clause 45.1 the contractor is liable to pay an assessed cost – whether or not that proves to be the actual cost. And, since there is no reference to incurred cost, the contractor appears to be liable to pay the assessed cost whether or not the employer actually incurs any cost.

Note in particular here the change of wording in clause 45.1 from that of clause 40.6 which does refer to the project manager assessing the cost 'incurred' by the employer.

The implications of this for the contractor are quite alarming. In the most extreme case he could be liable for the estimated replacement cost of the whole of the works for some minor infringement of the works information – such as building the works marginally, but inconsequentially, out of position.

At common law, following the *Ruxley Electronics* decision mentioned in section 9.7 above, such a liability no longer exists (if it ever did) and it is surprising to find the liability in a contract as progressive as the NEC.

In the event of a dispute on the matter it is doubtful if the adjudication procedure of the NEC would be of any assistance to the contractor; the employer would simply be enforcing the terms of the contract. But relief could perhaps be gained against oppressive use of the clause by reliance on clause 10.1 of the contract, which requires the employer to act in a spirit of mutual trust.

# Chapter 10

# Payments

## 10.1 Introduction

Section 5 is the main body of core clauses of the NEC dealing with payments. Its provisions are grouped under three headings:

- clause 50 – assessing the amount due
- clause 51 – payment
- clause 52 – actual cost.

But these clauses are only part of the payment scheme of the NEC. Much of the detail is found in the core clauses applicable to particular main options, including:

- clause 11 – defined terms for:
  - the prices
  - the price for work done to date
  - actual cost
  - disallowed cost
- clause 53 – the contractor's share
- clause 54 – the activity schedule
- clause 55 – the bill of quantities.

Also relevant, if included in the contract, are certain secondary options:

- Option J – advanced payments
- Option K – multiple currencies
- Option N – price adjustment for inflation
- Option P – retention.

In addition to the contract clauses the following entries in the contract data have to be noted:

- in part one provided by the employer:
  - the starting date
  - the currency of the contract

## 10.1 Introduction

  - the assessment interval
  - the interest rate
  - the payment period (if not three weeks)
  - the method of measurement (for Option B)
  - share percentages (for Option C or D)
  - exchange rates (for Option C, D, E or F)
  - advance payment details (for Option J)
  - currency details (for Option K)
  - indices details (for Option N)
  - retention details (for Option P).

- in part two, provided by the contractor:
  - the activity schedule (for Option A or C)
  - the bill of quantities (for Option B or D)
  - the tendered total of prices (for Options A, B, C or D)
  - cost component details.

**Essentials of the payment scheme**

Although there are different payment mechanisms for each of the six main options, some common characteristics can be extracted which define the essentials of the NEC payment scheme. These are:

- assessments of amounts due are made at not more than five week intervals
- certification is within one week of each assessment date
- payment is due within three weeks of each assessment date (unless stated otherwise)
- interest is due on late certification, under-certification or late payment.

**Amounts due**

The rules for calculating amounts due for both interim and final payments vary according to the main option used. In short, excluding specific adjustments, the amounts due are as follows:

Option A – lump sum contract
- interim amounts – the total of prices for completed activities
- final amount – the total of the prices of the activities.

Option B – remeasurement contract
- interim amounts – the quantities of completed work at bill of quantity rates and proportions of any lump sums
- final amount – remeasured value of the work in accordance with the bill of quantities.

## 10.1 Introduction

Option C – target contract with activity schedule
- interim amounts – actual cost paid plus fee
- final amount – tendered price as the activity schedule plus or minus the contractor's share.

Option D – target contract with bill of quantities
- interim amounts – actual cost paid plus fee
- final amount – remeasured value of the work in accordance with the bill of quantities plus or minus the contractor's share.

Option E – cost reimbursable contract
- interim amounts – actual cost paid plus fee
- final amount – actual cost paid plus fee.

Option F – management contract
- interim amounts – actual cost accepted for payment plus fee
- final amount – actual cost paid plus fee.

**Peculiarities of the payment scheme**

Although much of the NEC payment scheme is conventional, there are some peculiarities which should be noted.

Perhaps the most important of these from the employer's viewpoint is that the burden of assessing the amount due falls squarely on the project manager, and he carries this burden whether or not the contractor submits an application for payment (which he is not obliged to do). For Option A, assessment of the amount due is a straightforward matter of deciding which activities have been completed and the project manager's burden is comparatively light. For the other main options, however, particularly those where interim payments are based on costs paid or incurred, the assessment process can be time consuming and complex. And since the project manager is only allowed one week (clause 51.1) to make his assessment, he is likely to need plentiful support staff or the services of a professional quantity surveying firm. These expenses fall directly on the employer.

From the contractor's viewpoint an important peculiarity is, perhaps, that with main options C, D and E interim payments are made only on the basis of costs actually paid by the contractor. The cashflow and financing implications of this on the contractor are obvious and the probability is that many contractors will be reluctant to take on sizeable contracts under Option C, D or E without some change in the interim payment scheme. Employers will also have to be careful when using Options C, D or E to ensure that they have selected contractors of good financial standing.

Other peculiarities to note are:

- retention is not automatic – it applies only when secondary option P is incorporated
- interim payments suffer a 25% deduction until a first programme is submitted for acceptance
- there is no provision to stipulate a minimum amount for interim payments
- there is no reference to a final certificate, nor is any certificate given special contractual status.

## 10.2 Assessing the amount due

### Clause 50.1 – assessment procedure

Clause 50.1 commences by requiring the project manager to assess the amount due at each assessment date. It then continues by stating when assessment dates occur.

The significance of the obligation on the project manager to assess the amount due has been discussed above. However, note that by clause 50.4 if the contractor does make an application for payment the project manager has to consider it in his assessment. Note also that by clause 51.1 the project manager is required to issue his certificate within one week of the assessment date.

The first assessment date is decided by the project manager 'to suit the procedures of the Parties' but it should not be later than the assessment interval after the starting date. In the contract data (but not in the written conditions) it is indicated that the assessment interval should be not more than five weeks, so the first assessment date should never be more than five weeks after the starting date (a date fixed by the employer in the contract data).

The phrase 'to suit the procedures of the Parties' suggests that the project manager should take into account the views of both parties, but sometimes, of course, there are differences, with the employer concerned with regularising payment dates and the contractor concerned with internal management accounting dates.

Later assessment dates occur:

- periodically – at the end of each assessment interval – but only until completion of the whole of the works
- singularly – at completion of the whole of the works
- singularly – four weeks after the issue of the defects certificate (of which there is only one)
- exceptionally – after completion of the whole of the works when:
  ○ an amount due is corrected
  ○ a payment is made late.

## 10.2 Assessing the amount due

One effect of this scheme is that if the contractor does work after completion he may have to wait up to twelve months for payment. This may well be encouragement to the contractor not to leave work outstanding at completion, but it will be a harsh rule if applied to payments for repair or emergency work undertaken after completion which is at the employer's expense (see clause 82.1).

The assessment after the issue of the defects certificate is probably intended to be the assessment which will lead to what is normally called the final certificate. The involvement of the contractor in this is not specifically stated but applying clause 50.4 suggests that if the contractor wants to have his say in the final assessment he must put in his final account before the last assessment date.

Amounts which become due as a result of adjudication can, it would seem, be treated as corrections and certification after completion takes place as and when necessary.

### Clause 50.2 – the amount due

Clause 50.2 describes in general terms the amount due. It is:

- the price for work done to date, *plus*
- other amounts to be paid to the contractor, *less*
- amounts to be paid by or retained from the contractor.

The clause confirms that any value added tax or sales tax applicable by law 'is included in the amount due'.

The wording of the latter provision is a little unfortunate because on one interpretation it could mean that the contractor's prices are inclusive of VAT; whereas what it presumably is intended to mean is that the prices are exclusive of VAT. The phrase 'is included' in the last sentence of the clause is better read as 'is to be included'.

But even given that meaning there is still an administrative problem to overcome on value added tax in that, unless the employer operates an approved self billing system, value added tax only becomes due on receipt of a VAT invoice. In practice therefore it will be normal for the project manager and the contractor to liaise so that the contractor's certificate is accompanied by a matching VAT invoice.

For comment on the detail of the amounts due under the various main options, see sections 10.5 to 10.10 below.

### Clause 50.3 – failure to submit programme

Clause 50.3 is an unusual provision which penalises the contractor if he delays in submitting a first programme for acceptance. The clause provides that:

- if no programme is identified in the contract data, *then*
- one quarter of the price for work done to date is retained in assessments of amounts due, *until*
- the contractor has submitted a first programme for acceptance showing all the information required by the contract.

Note that retention applies only to the price for work done to date and not to other components of the amount due. Note also that the retention is made at the assessment stage by the project manager so the certificate itself is for the reduced amount.

One aspect of the clause which has caused some concern is whether it has the potential to be abused to reduce the employer's payment liabilities. The wording of the clause appears to recognise the possibility by allowing the retention only until the first programme is submitted – not until it is accepted. But the clause is silent on the not improbable situation that the contractor and the project manager may be in dispute on whether the submitted programme shows all the information which the contract requires. In such circumstances it seems to be open to the project manager to make the deduction and to the contractor to seek adjudication if dissatisfied.

Some commentators have suggested that the clause may be challengeable in law as a penalty clause and that it may not be enforceable. As far as interim payments are concerned it is questionable whether the argument would succeed because the retention is an agreed precedent to a contractual entitlement rather than a form of liquidated damages for breach. However, if a situation arose where a complying first programme was never submitted and the clause was used (as appears possible) to reduce the final contract price by 25%, that would surely be a penalty.

Finally, it is worth noting that clause 50.3 does not necessarily lead to an accepted programme coming into place early in the contract. Clause 31.3 states four reasons for not accepting a programme but clause 50.3 refers to only one – lack of information. So the motivation for submitting a programme which is acceptable on the other grounds is not to be found in clause 50.3. For that the contract relies on clause 64.2 – the assessment of compensation events.

### Clause 50.4 – application by the contractor

Clause 50.4 contains two useful provisions. Firstly, if the contractor does submit an application for payment on or before the assessment date the project manager is required to consider the application when making his assessment. Secondly, the project manager is to give the contractor details of how an amount due has been assessed.

Clearly if the contractor only submits his application on the assessment date itself that leaves the project manager with very little time (one week under clause 51.1) to consider it. But for commercial reasons (maximising the amount due) and contractual reasons (maximising the potential for interest) the contractor is likely to submit his application as late as permissible.

The requirement for the project manager to give details of his assessment is, it is suggested, applicable whether or not the contractor has made an application. But to avoid any uncertainty it might have been better if this requirement had been put in clause 50.1 or clause 50.5. It would also have been better if some timescale had been attached to the requirement.

**Clause 50.5 – corrections of assessments**

Clause 50.5 states only that the project manager corrects any wrongly assessed amount due in a later certificate. But the clause is not, perhaps, as simple as it seems.

The question is, who decides whether an amount has been wrongly assessed? Obviously there is no problem if the wrong assessment is confirmed by an adjudicator, but short of involving an adjudicator it is none too obvious how the reality of a wrong assessment can be established or how the project manager can be obliged to admit to making a wrong assessment. In most cases wrong assessments, or alleged wrong assessments, will automatically be corrected in later certificates without there being any admission of an earlier wrong assessment. So if the purpose of the clause is to establish the contractor's right to interest under clause 51.3 it may not be too effective.

On a point of detail note that any wrongly assessed amount becomes due 'in a later payment certificate'. The contractor does not appear to have an entitlement (other than interest) to correction in the 'next' certificate.

## 10.3 Payments

**Clause 51.1 – certification and obligations**

The provisions of clause 51.1 can be summarised as follows:

- the project manager is required to certify payment within one week of each assessment date
- the first payment is the whole amount certified (subject to any retentions)

- subsequent payments are changes in amounts due from certificate to certificate
- if the change is a reduction the contractor pays the employer
- when the change is an increase the employer pays the contractor
- payments are made in the currency of the contract (as specified in the contract data) unless otherwise stated (i.e. Option K on multiple currencies applies).

Note that strictly the project manager does not certify amounts due (except in the first certificate) but only changes in amounts due. The terminology may lead to some confusion but no doubt most project managers will issue certificates showing both amounts due and changes in amounts due.

The requirement for the contractor to pay the employer for any change reducing the amount due is proper in principle but its application in the NEC is arguably more onerous than normal. This is because by clause 51.2 the contractor must make the payment within three weeks or pay interest on late payments, whereas most contracts leave overpayments to be settled in the final certificate.

**Clause 51.2 – time for payment and late payment**

The first provision of clause 51.2 requires the employer to pay on each certificate within three weeks of the assessment date or within such other time as may be stated in the contract data.

As the project manager is allowed one week from the assessment date to certify (and will probably need all of that), the employer has effectively two weeks under the standard time from certification to payment. Note that if the project manager over-runs his time in certifying, that cuts into the employer's time since the payment time is linked to the assessment date and not to the date of certification.

Having fixed the time for payment, clause 51.2 goes on to provide for interest on late payment. Interest is calculated for the period between the due payment date and the actual payment date and is to be included in the assessment following the late payment. Clearly interest is payable without an application by the entitled party and it is up to the project manager to ensure that his assessments include for interest when appropriate. The interest rate is that fixed by the employer in part one of the contract data, which is to be a rate not less than 2% above a specified bank rate. Clause 51.5 confirms that interest is compounded annually.

**Clause 51.3 – interest on corrected amounts**

Clause 51.3 provides for interest on amounts undercertified and corrected in later certificates. The clause appears to be unusually strict in its application because it states that interest is paid on the correcting amount if a certificate is corrected either:

- by the project manager in relation to
  - a mistake, or
  - a compensation event, *or*
- following a decision of an adjudicator or disputes tribunal.

Note, in relation to the reference in the clause to a 'mistake', that this is not confined to a mistake by the project manager. It apparently extends to mistakes by the contractor. However, since it is the responsibility of the project manager to assess amounts due on certificates without any obligatory input from the contractor, any correction to a certificate will appear at face value to arise from a mistake by the project manager – whether by a mistake of law or a mistake of fact.

There may occasionally be circumstances where the contractor has obstructed the project manager from making a correct assessment, perhaps by the non-disclosure of records or non-compliance with the accepted programme, and where the contractor cannot argue that the circumstances constitute a mistake, but otherwise it is difficult to see how corrections can avoid attracting interest.

Deliberate under-certification does perhaps avoid the scope of clause 51.2 but that would seem to be a breach of clause 50.1 entitling the contractor to a compensation event – clause 60.1(18).

The intentions of the reference in clause 51.3 to corrections for compensation events are not wholly clear. It may simply be intended that if compensation events are revalued then interest accrues from the original valuation. But the intention may be much wider such that interest runs from the first certification date after a compensation event is claimed until certification of the proper value. This would be not far removed from giving the contractor an automatic right to financing charges on claims.

Interest under clause 51.3 is calculated as for clause 51.2 and it runs from the date of the incorrect certificate to the date of the corrected certificate.

**Clause 51.4 – failure to certify**

Clause 51.4 comes into operation when the project manager does not issue a certificate which he should issue. Interest is then payable on the amount he should have certified from the due date until the actual date of certification.

The clause is necessary because clause 51.1 relates payment obligations to certificates, so if there is no certificate there is no obligation to pay and no breach of contract. However, as worded, the clause has the unusual effect that the employer may become liable for interest even though he pays within the time stipulated by clause 51.2. This will happen if the project manager is late by up to two weeks in certifying but the employer still pays within the standard three weeks of the assessment date.

Even more unusual, if the project manager fails to certify by the due payment date then it appears that the contractor becomes entitled to double interest – interest under clause 51.2 for late payment and interest under clause 51.4 for late certification. In both cases, of course, the interest flow can be reversed if amounts are due from the contractor to the employer. This could lead to the extraordinary result that the contractor could become liable for interest to the employer as a result of the project manager's breach of duty.

**Clause 51.5 – rate of interest**

The interest rate applicable throughout the NEC is an identified term fixed by the employer in part one of the contract data. The data entry stipulates (but no contract clause does so) that the interest rate shall be not less than 2% above a specified bank rate.

Clause 51.5 states that interest is to be calculated at the identified rate and is to be compounded annually. This compares unfavourably with other contracts – the 1995 IChemE Red Book, for example, states that interest accrues on a daily basis – and it also compares unfavourably with commercial realities.

## 10.4 Actual cost

Actual cost is a defined term found in:

- clause 11.2(26) for Option F
- clause 11.2(27) for Options C, D and E
- clause 11.2(28) for Options A and B

Clause 52.1 states certain matters common to all the above.

**Clause 52.1**

The first provision of clause 52.1 states that all the contractor's costs not included within actual cost (as defined) are deemed to be included in the fee percentage. In other words the contractor must allow in the fee percentage which he tenders in part two of the contract data, for all costs not expressly covered within the definition of actual cost.

## 10.4 Actual cost

The second provision of clause 52.1 is that all amounts included in actual cost are to be:

- at open market or competitively tendered prices
- with all discounts, rebates and taxes which can be recovered deducted.

This is some protection to the employer that payments made on an actual cost basis will not be excessive – at least where the contractor's costs are to be supported by subcontractor or supplier invoices.

The main impact of clause 52.1 is, of course, on the cost reimbursable main options but it does apply to all the options in the assessment of compensation events.

## 10.5 Payments – main option A

The following core clauses are particular to main option A – the priced contract with activity schedule:

- clause 11.2(20) – definition of the prices
- clause 11.2(24) – definition of price for work done to date
- clause 11.2(28) – definition of actual cost
- clause 54.1      – the activity schedule
- clause 54.2      – revision of activity schedule
- clause 54.3      – reasons for not accepting the activity schedule.

### Clause 11.2(20) – definition of the prices

The prices are defined as:

- the lump sum prices
- for each of the activities in the activity schedule
- unless changed later in accordance with the contract.

The definition suggests that a contract let under main option A is a contract for a collection of various activities with individual lump sum prices, rather than a straightforward lump sum price contract. For comment on this see Chapter 2, section 3.

The phrase 'unless later changed' ensures that the definition of the prices remains valid from tender to completion. But note that the changes are to be 'in accordance with this contract' and they do not extend to other changes of the lump sum price or prices which may be agreed between the parties.

### Clause 11.2(24) – the price for work done to date

The defined term 'the Price for Work Done to Date' governs the amount the contractor is entitled to be paid (clause 50.2) at both interim and final stages. The definition for the price for work done to date is the total of the prices:

- for each group of completed activities, *and*
- each completed activity which is not in a group
- which is without defects which would delay following work or which would be covered immediately by following work.

It is up to the contractor how he forms his activity schedule and whether or not he shows grouping of activities. The contractor has to keep in mind that only completed groups of activities or completed single activities attract the right to interim payment. There is no contractual advantage to the contractor in grouping activities together, but the disadvantage is obvious enough.

The reference to defects in the price for work done to date is worded so as to permit a certain level of defects to be tolerated without prejudicing the contractor's right to payment.

Note that activities which are not included in the activity schedule do not come within the scope of the definition of 'Prices' in clause 11.2(20) and therefore are not within the scope of 'the Price for Work Done to Date' in clause 11.2(24). Consequently omitted activities never attract a right to payment and are deemed to be covered in the listed activities.

### Clause 11.2(28) – actual cost

Actual cost for Option A is defined as:

- the cost of components in the schedule of cost components
- whether subcontracted or not
- excluding the cost of preparing quotations for compensation events.

The first part of the definition restricts recovery on cost basis to the items detailed in the schedule of cost components.

The use of capital initials for the term schedule of cost components suggests that it is a defined term of the NEC (which it is not) but it does serve to distinguish between the two schedules of cost components. Under Option A, the shorter schedule can only be used for assessing actual cost if clause 63.11 on the assessment of compensation events applies.

It is, of course, the case that in Option A the only use of the actual cost provisions is in the assessment of compensation events, so in practice actual cost will frequently be determined under the shorter schedule.

The second part of clause 11.2(28) referring to subcontracted work prevents subcontract invoices being put forward as evidence of actual cost (at least under this option) and requires subcontractor costs to be calculated with reference to the schedule of cost components. For the main contractor one potentially unfortunate aspect of this is that the subcontractor's overheads are deemed to be included in the contractor's fee percentage; another is that the contractor's financial liability to his subcontractor may be greater than the contractor's financial recovery under the main contract.

The final part of clause 11.2(28), stating that actual cost excludes the cost of preparing quotations for compensation events, is a surprising provision to find in the NEC, which is promoted as being a fair contract.

Since most compensation events arise from defaults for which the employer is responsible, and some are expressly breach of contract provisions, the contractor should be able to recover the costs he incurs. And where the contractor is instructed to prepare quotations or alternative quotations for proposed changes which are not later instructed, the case for payment of the contractor's costs is so strong that arguably it must escape the exclusion in clause 11.2(28) to allow preparation of the quotation to be treated as a compensation event in its own right.

**The activity schedule**

For general comment on the activity schedule see Chapter 2, section 3.

**Clause 54.1 – information in the activity schedule**

The purpose of clause 54.1, which states only that information in the activity schedule is not works information or site information, is to ensure that the activity schedule does not acquire unintended contractual effect in respect of the obligations of the parties.

**Clause 54.2 – changes to the activity schedule**

In order to maintain the integrity of the payment system for Option A, which relies on the identification of completed activities, and the integrity of the scheme for assessment of compensation events, the NEC needs compatibility between the activity schedule and the accepted programme.

Clause 54.2 requires the contractor to submit a revised activity schedule to the project manager for acceptance when:

- he changes a planned method of working
- at his discretion
- so that the activity schedule does not comply with the accepted programme.

It is not wholly clear why clause 54.2 is confined to changes in methods of working made at the contractor's 'discretion'. Changes imposed on the contractor by compensation events are just as likely to affect the activity schedule. Perhaps the explanation is that clause 54.2 is to be seen as dealing only with mandatory revisions and that it does not exclude the contractor's right to make changes to his activity schedule for matters beyond his control.

And obviously when there are delaying compensation events which are affecting the contractor's cashflow, the contractor should have the right to change his activity schedule notwithstanding his entitlement to recover in the course of time financing charges as an element of cost under the schedules of cost components.

**Clause 54.3 – reasons for not accepting a revised activity schedule**

Clause 54.3 states three reasons for the project manager not accepting a revision of the activity schedule:

- the revision does not comply with the accepted programme
- the changed prices are not distributed reasonably between the prices
- the total of the prices is changed.

This final reason provides confirmation, if it is needed, that the total of the prices in the original activity schedule is the total contract price (subject, of course, to contractual adjustments).

The consequences of not accepting a revised activity schedule are to leave payments and compensation events to be valued in accordance with the previous activity schedule.

## 10.6 Payments – main option B

The core clauses particular to main option B – the priced contract with bill of quantities – are:

- clause 11.2(21) – definition of the prices
- clause 11.2(25) – definition of the price for work done to date
- clause 11.2(28) – definition of actual cost
- clause 55.1 – the bill of quantities.

## Clause 11.2(21) – definition of the prices

The prices are defined as:

- the lump sum
- the amounts obtained by multiplying the rates by the quantities for the items in the bill of quantities
- unless changed in accordance with the contract.

The traditional approach in bill of quantities contracts is to refer to rates and prices – with rates applying to remeasured items and prices to lump sums.

For reasons which are not obvious the NEC takes an unusual approach and defines prices so that a rate times a quantity is regarded as a price. The contractual implications of this may take some time to emerge.

## Clause 11.2(25) – the price for work done to date

The price for work done to date is stated in clause 11.2(25). It is:

- the quantity of completed work for each bill of quantities item multiplied by the appropriate rate, *and*
- such proportion of each lump sum in the bill of quantities as is completed.

The clause states further that completed work 'in this clause' means work without defects which would either delay or be covered by immediately following work. The phrase 'in this clause' is presumably intended to emphasise that 'completed work' as mentioned in the clause for payment purposes is not to be taken as completed work for other contractual purposes.

## Clause 11.2(28) – actual cost

This is the same clause as used for Option A; for comment see the section 5 above.

## Clause 55.1 – the bill of quantities

Clause 55.1 is of similar purpose to clause 54.1 and it states only that information in the bill of quantities is not works information or site information.

## 10.7 Payments – main option C

The following core clauses are particular to main option C – the target contract with activity schedule:

- clause 11.2(20)        – definition of the prices
- clause 11.2(23)        – definition of the price for work done to date
- clause 11.2(27)        – definition of actual cost
- clause 11.2(30)        – definition of disallowed cost
- clause 50.6            – assessing the amount due
- clauses 52.2 and 52.3  – records of actual cost
- clauses 53.1 to 53.5   – the contractor's share
- clauses 54.1 to 54.3   – the activity schedule.

**Clause 11.2(20) – definition of the prices**

This is the same clause as for main option A; for general comment see section 5 above.

However, it is important to note that the prices have far less contractual importance in Option C than in Option A. In particular, in Option A the prices form the foundation of the price for work done to date (which governs the final contract price), whereas in Option C the prices play no part in the price for work done to date and only affect the final contract price through calculations in the contractor's share (clause 53).

**Clause 11.2(23) – definition of the price for work done to date**

Comment on clause 11.2(23), which states that the price for work done to date is the actual cost which the contractor has paid plus the fee, is given in section 10.1 above.

It is known that some users of Option C are not adopting this definition in practice and are paying the contractor for costs incurred rather than costs paid.

**Clause 11.2(27) – definition of actual cost**

Clause 11.2(27) defines actual cost as:

- the amount of payments due to subcontractors
- the cost of components in the schedule of cost components for work not subcontracted
- less any disallowed cost.

The principal difference between this clause, which applies to main options C, D and E, and clause 11.2(28) applying to main options A and B, is that subcontractors' invoices are a recognised part of actual cost. This reflects the cost reimbursable aspects of Options C, D and E.

Other differences between clauses 11.2(27) and 11.2(28) are that clause 11.2(27) does not expressly exclude the costs of preparing quotations for compensation events, but it does expressly exclude disallowed cost.

In Option C, the contractor can recover the costs of preparing quotations subject only to the effect such costs may have on the contractor's share. Disallowed costs have to be defined in cost reimbursable contracts, otherwise the employer is liable to pay for remedying the contractor's defaults.

**Clause 11.2(30) – disallowed cost**

Disallowed costs under clause 11.2(30) fall into four groupings:

- costs not substantiated
- costs overpaid
- costs relating to the correction of defects
- costs not related to providing the works.

The clause lists disallowed costs as follows:

- costs not justified by accounts and records
- costs which should not have been paid to subcontractors
- costs incurred because the contractor did not follow an acceptance or procedure in the works information or did not give an early warning
- costs of overpaying subcontractors in respect of compensation events
- costs of correcting defects after completion
- costs of correcting defects caused by not complying with requirements in the works information
- plant and materials not used to provide the works – after allowing for reasonable wastage
- resources not used to provide the works or not taken away when requested.

Most of these are straightforward in principle even if open to argument on their application in particular circumstances. However, one which takes some understanding is that relating to defects before completion. The exact wording is 'correcting Defects caused by the Contractor not complying with a requirement for how he is to Provide the Works stated in the Works Information'.

This wording, taken with the definition of a defect in clause 11.2(15), indicates that not all costs of correcting defects are disallowed. Defects caused by the contractor's design can apparently be corrected at cost – which raises questions on the suitability of using Option C (and other cost reimbursable options) with contractor's design. And presumably there is intended to be some significance in the words 'not complying with a requirement', so it may be that certain other defects can be corrected at cost.

### Clause 50.6 – assessing the amount due

Clause 50.6 deals with the situation when payments for actual cost are made by the contractor in a currency other than the currency of the contract. The clause provides that in such circumstances the amount due to the contractor is calculated by reference to the currency of the cost, but for calculation of the fee and the contractor's share payments it is converted to the currency of the contract.

### Clause 52.2 – records of actual cost

In keeping with the cost reimbursable aspect of Option C, clause 52.2 requires the contractor to keep records of his actual cost including, and expressly:

- accounts of payments
- records showing payments made
- records relating to compensation events for subcontractors
- other records and accounts as stated in the works information.

### Clause 52.3 – inspection of records

Clause 52.3 provides that the contractor shall allow the project manager to inspect the records at any time within working hours. Since it is the project manager's obligation to assess amounts due to the contractor this is clearly an essential provision.

### The contractor's share

For any contractor considering entering into a target cost contract, one of the key questions is what are the potential risks and rewards arising from excesses or savings on the target cost. For general comment on this see Chapter 2, section 5.

## 10.7 Payments – main option C

Some target cost contracts have very simple formulae for fixing the contractor's and the employer's shares of excesses and savings. The NEC has an incremental scheme which requires the employer to state in the contract data (in percentage terms) ranges of deviation from the target cost and corresponding share percentages. In arithmetical terms the scheme is quite straightforward, but expressing it in words appears to have created problems.

### Clause 53.1 – calculating the share

Construction Industry Law Letter offered a prize for (amongst other things) the decoding of this clause. It is best understood by applying sample numbers. See, for example, page 54 of the NEC:ECC Guidance Notes.

### Clause 53.2 – payment of the share

This clause appears to do little more than state the obvious, namely, that the contractor is paid his share of any saving and that he pays his share of any excess. But the clause may be necessary since in the event of excess the contractor will have been overpaid and the obligation to repay his share of the excess needs to be clearly stated.

### Clause 53.3 – first assessment of the share

Clause 53.3 confirms that the first assessment of the contractor's share is not made until completion of the whole of the works. The assessment is made using the project manager's forecasts for the final price for the work done to date and the final prices (which are the tender prices adjusted for compensation events and the like).

The clause requires the contractor's share to be included in the amount due 'following' completion of the whole of the works. This probably means that the contractor's share is included in the assessment made 'at' completion under clause 50.1.

Because the NEC does not provide for interim assessments of the contractor's share before completion, the contractor is fully reimbursed on a cost basis up to completion. In this respect the NEC is significantly different from target contracts where the target mechanism operates as a guaranteed maximum cost and where payment cuts off when this is reached.

### Clause 53.4 – final assessment of the share

The final assessment of the share is made using the final price for work done to date and the final total of the prices. This share is included in the final amount due, so unless there are complications, it should be included in the amount certified after the issue of the defects certificate.

### Clause 53.5 – proposals for reducing actual cost

Clause 53.5 is an important provision included to encourage the contractor to maximise his share potential. It provides that the prices are not reduced when the project manager accepts a proposal by the contractor changing the works information 'provided by the Employer' so that actual cost is reduced. In other words if the contractor can identify changes or variations resulting in savings in the employer's design the contractor gets some benefit from the savings by way of increased share.

Note that the clause only applies to works information 'provided by the Employer'. This is because a change in works information provided by the employer is a compensation event (clause 60.1(1)) which would normally lead to a change in the prices, whereas a change in works information provided by the contractor at his own request is expressly not a compensation event (clause 60.1(1)), and the contractor is, therefore, automatically entitled to his share of any saving.

### The activity schedule

In Option C, the activity schedule serves a different purpose than in Option A. Whereas in Option A the activity schedule is used to assess interim payments, in Option C the activity schedule is used only in the assessment of compensation events and in detailing the total of the prices for the calculation of the contractor's share.

### Clauses 54.1 to 54.3 – activity schedule

These are the same clauses as used in Option A – see section 10.5 above for comment.

## 10.8 Payments – main option D

The core clauses particular to main option D – the target contract with bill of quantities – are:

## 10.8 Payments – main option D

- clause 11.2(21) – definition of the prices
- clause 11.2(23) – definition of the price for work done to date
- clause 11.2(27) – definition of actual cost
- clause 11.2(30) – definition of disallowed cost
- clause 50.6 – assessing the amount due
- clauses 52.2 and 52.3 – records of actual cost
- clauses 53.1 to 53.5 – the contractor's share
- clause 55.1 – the bill of quantities.

Clauses 11.2(21) and 55.1 above are common to Option B – for comment see section 10.6 above. The remaining clauses are common to Option C – for comment see the previous section.

The bill of quantities in Option D serves the same purpose as the activity schedule in Option C and it is not used for assessing interim payments.

## 10.9 Payments – main option E

The following core clauses are particular to main option E – the cost reimbursable contract:

- clause 11.2(19) – definition of the prices
- clause 11.2(23) – definition of the price for work done to date
- clause 11.2(27) – definition of actual cost
- clause 11.2(30) – definition of disallowed cost
- clause 50.7 – assessing the amount due
- clauses 52.2 and 52.3 – records of actual cost.

Because main option E is a fully cost reimbursable contract it omits clauses 53.1 and 53.5 relating to the contractor's share and, of course, any reference to either an activity schedule or a bill of quantities.

Clauses 11.2(23), 11.2(27), 11.2(30), 52.2 and 52.3 common to Options C and D (and also F), have been considered earlier.

Clauses 11.2(19) and 50.7 are common only to Options E and F.

### Clause 11.2(19) – definition of the prices

The prices are defined simply as the actual cost plus the fee. There is no significance in the use of the term 'Prices' as opposed to 'Price'. It is used simply to achieve compatibility with other clauses of the contract.

### Clause 50.7 – assessing the amount due

Clause 50.7 is similar to clause 50.6 used in Options C and D for dealing with payments made by the contractor in currencies other than the currency of the contract, except that it omits the reference in clause 50.6 to the contractor's share.

## 10.10 Payments – main option F

The core clauses particular to main option F – the management contract – are:

- clause 11.2(19) — definition of the prices
- clause 11.2(22) — definition of the price for work done to date
- clause 11.2(26) — definition of actual cost
- clause 11.2(29) — definition of disallowed cost
- clause 50.7 — assessing the amount due
- clauses 52.2 and 52.3 — records of actual cost.

From the above list only clauses 11.2(22), 11.2(26) and 11.2(29) differ from the clauses also applicable to main option E.

### Clause 11.2(22) – definition of the price for work done to date

The definition here differs from that used for Options C, D and E in that the price for work done to date is defined as the amount the contractor 'has accepted for payment' plus the fee – and not as the amount 'paid'.

### Clause 11.2(26) – definition of actual cost

Actual cost is defined as:

- the amount of payments due to subcontractors
- for work which the contractor is required to subcontract
- less any disallowed cost.

This definition does not appear to be compatible with clause 20.2 in that it assumes all the work is done by subcontractors and it makes no provision for any work the contractor does himself.

### Clause 11.2(29) – definition of disallowed cost

The difference between the definition here of actual cost and the definition in clause 11.2(30) applicable to Options C, D and E, is that clause 11.2(29) omits the references in clause 11.2(30) to:

## 10.10 Payments – main option F

- the costs of correcting defects
- the costs of plant and materials not used to provide the works
- the costs of resources not used to provide the works or not taken away when requested.

This suggests that there is less disallowed cost under Option F than under the other cost reimbursable options, but that is not intended to be the case. Normally all the work in Option F will be subcontracted as packages on lump sum prices and each subcontractor will be responsible for correcting his own defects at his own cost. Only if the contractor does some of the work himself will the omitted items of disallowed costs be of any significance.

# Chapter 11

# Compensation events

## 11.1 Introduction

The compensation event procedure of the NEC is one of its more radical departures from traditional principles. It has attracted glowing support from some quarters and stinging criticism from others. Designed to encourage good management and to reduce conflict it is, like much of the NEC, easier to promote by reference to its ideals than by reference to its wording.

The theory of the procedure as explained by the promoters of the NEC is that the contractor is compensated on an actual cost basis for variations and claims. And accordingly, it is said, because he carries no additional financial risk, he is indifferent to decisions made by the project manager and is motivated to minimise time and cost overruns. In reality, however, the procedure offers no guarantee to the contractor that he will recover the actual costs he incurs. And the risks he carries are arguably greater than those he carries under traditional contracts. Nevertheless, with astute management of the procedure the contractor may be able to turn it to his financial advantage.

As for employers, they may view the compensation event procedure as a major obstacle to any aspirations they hold for price certainty in their contracts in that the NEC, by detailing a list of entitlements to additional payments the like of which is not seen in any other standard form of contract, gives apparent encouragement to claims rather than discouragement.

**Application of the procedure**

One aspect of the procedure which must give cause for concern to those involved in its use is the likely frequency of its application. The procedure appears to have been drafted in the belief that a compensation event is a rare event; and the complexity and timing arrangements of its provisions are not really suited to anything else. In fact the great majority of the permitted changes to the contract price under the NEC fall under the compensation event procedure, including the valuation of instructions with the most trivial of cost implications. There is no mechanism in the contract allowing for an 'extras' account with 'star' rates or for by-passing the compensation event procedure where the costs of administration would exceed, or be an unduly high proportion of, the amount claimed. The result is that compensation events are not rare events but everyday events. Reports from early users of the NEC are of hundreds of compensation events on comparatively straightforward schemes.

Not surprisingly many of the reports confirm either a breakdown of the compensation event procedure or an agreement between the parties that its administrative provisions should be greatly relaxed and that it should not apply to items below a particular figure. Even so, the demands on the contractor's staff and the project manager's staff are high and parties entering into NEC contracts need to recognise the need for additional resources.

This will obviously increase costs during construction but the hope is that it will reduce costs in dealing with tail-end claims.

**Programme involvement**

One characteristic of the compensation event procedure which exacerbates its administrative burdens is that it covers both time and money. The procedure relies on programmes for the assessment of actual cost and it requires a revised programme for any event involving delay or disruption.

To cope with this, contractors are finding it necessary to have on-site programmers and estimators as well as the usual quantity surveyors. And in some cases, where the programme is computerised, contractors have established facilities on site to issue regular, if not daily, revisions of the programme. Unless project managers are able to match this degree of efficiency and respond to revised programmes within the two weeks they are allowed by clause 31.3, and to quotations within the two weeks allowed by clause 62.3, they risk losing control over time and cost.

**Outline of the procedure**

In summary the compensation event procedure works as follows:

- Certain events are stated in core clauses 60.1 to 60.6 to be compensation events. Other events are stated in secondary options J, T and U to be compensation events. Additional events may also be stated by the employer in the contract data (but see the note on these in section 11.2 below).
- The project manager is required to notify the contractor of compensa-

tion events arising from instructions; and the contractor is required to notify the project manager within two weeks of other events occurring or events not notified by the project manager.
- The contractor is required to submit quotations for compensation events showing both time and money implications. A revised programme must be submitted with the quotation if there are time or disruption implications.
- Quotations for changes to the contract price are based on assessments of actual cost incurred or forecast to be incurred. But note that actual cost is 'actual cost' within the meaning of the contract and as determined from the schedules of cost components. It is not true actual cost.
- The project manager is permitted to make his own assessment if the contractor does not submit his quotation on time or if the contractor's programme is not in order or if he (the project manager) decides that the contractor's assessment is incorrect.
- The project manager is required to notify the contractor when a quotation is accepted or his own assessment is made, and the contract price and the times for completion are changed accordingly.

**Defining a compensation event**

The NEC lists the events it states to be compensation events but it does not define what a compensation event is. The contractual function of a compensation event is left to be derived from the provisions of the contract.

Clause 62.2 is helpful. This states that quotations for compensation events comprise proposed changes to the prices and any delay to the completion date. Clause 65.1 adds to this by stating that the project manager implements each compensation event by notifying the contractor of the quotation he has accepted or his own assessment. So clearly a compensation event is an event which can lead to an increase in the contract price and/or an increase in the time for completion.

However, not all compensation events which occur necessarily lead to either. Clause 61.4 indicates that there is no change to the contract price or the time for completion if the compensation event arises from the fault of the contractor or if it has no effect on actual cost or completion.

It would seem, therefore, that a compensation event is a listed event which, if it occurs, allows for changes in the contract price or time for completion providing it does not arise from a fault of the contractor.

**Do compensation events exclude other remedies?**

A question of key importance in the analysis of the NEC is whether or not its compensation event procedure provides an exclusive and exhaustive

## 11.1 Introduction

scheme for dealing with claims. Or, put another way, does the procedure exclude the contractor's common law rights to damages for breach of contract?

There is no ready answer to this question. The law is clear enough in that the right to claim damages for breach of contract can only be excluded by clear words – see *Hancock* v. *Brazier* (1966). But where doubt arises is whether a contractual scheme which does not expressly exclude common law rights but merely provides an alternative remedy is to be taken as an exclusion. Lawyers seem to be divided on this, which suggests that there is no general rule which is decisive in every case. Perhaps it is a question which can only be decided on the construction of the whole of the contract in each particular case. However, note the recent case of *Milburn Services Ltd* v. *United Trading Group (UK) Ltd* (1995) where the absence of express words excluding the common law right to damages seems to have been an important factor in the decision to regard the contractual scheme as supplementary to, but not in substitution of, common law rights.

There can be little doubt that when the NEC was first produced its promoters thought or hoped that they had devised a scheme which would eliminate end of contract claims. That message was fairly well paraded. And when the promoters realised that the list of compensation events in the first edition of the NEC was by no means comprehensive of the employer's possible defaults, they revised the list for the second edition and included most significantly event number 18 which reads 'A breach of contract by the Employer which is not one of the other compensation events in this contract'.

So clearly the NEC compensation event procedure as it stands in the second edition is fully comprehensive in that it covers all possible breaches by the employer. But whether that will be enough to exclude common law rights remains debatable.

The points which go against the proposition that the NEC excludes the contractor's common law rights to claim damages for breach remain formidable:

- there is no express exclusion of common law rights
- the contractor's entitlements under the compensation event procedure are rigidly prescribed by timing and administrative requirements
- the amounts recoverable under the compensation event procedure can fall well short of the amounts due as damages for breach of contract
- the contractor is required to forecast the costs of the employer's breaches of contract and his recovery is limited to his forecast
- it is arguable that assessments of compensation events by the project manager are not open to review under the disputes resolution procedure and are final and binding on the parties
- the contract preserves the employer's common law rights.

One point which does need to be made here is that even if the contractor's common law rights to damages are preserved in the NEC as a matter of law, those rights are enforceable only in respect of breach of contract and not in respect of pure contractual entitlements. So, in so far that the list of compensation events contains both items of breach and items of entitlement (e.g. unforeseen ground conditions), any claims in respect of the latter can only be pursued under the compensation event procedure.

**Is the compensation event procedure fair to the contractor?**

One of the major surprises which comes out of analysis of the compensation event procedure is that far from excelling as a scheme which is fair to the contractor, it places more risks on the contractor than other standard forms. It is hard to believe that this was the intention, but the explanation seems to be that by combining the scheme for the valuation of variations with the scheme for the valuation of the employer's breaches of contract, those drafting the NEC have overlooked the inequity of putting risk from the employer's breaches on to the contractor.

The risks which cause most concern are:

- the contractor's risk of losing his rights under the notice requirements
- the contractor's risk of making an inadequate forecast
- the contractor's risk that the amount of assessed actual cost recovered will be less than true actual cost, a risk to which the contractor is highly vulnerable under main options A and B in respect of subcontractor costs.

**Unusual features of the compensation event procedure**

In addition to the matters discussed above there are two further points which deserve a brief mention in this introduction to the compensation events.

The first point comes out of clause 65.2 which states that the assessment of a compensation event is not revised if a forecast upon which it is based is shown by later recorded information to have been wrong.

One effect of this appears to be that the NEC adopts what is known as the first-in-line approach to causative events. Thus if there is a delaying event for which the contractor is entitled to be paid and that event is overtaken by a delaying event for which the contractor is responsible, the contractor continues to be paid for the first event even though he is no longer suffering loss as a result of that event.

The second point comes out of clause 63.3. This states that a delay to the completion date is assessed as the length of time that planned completion

*11.1 Introduction* 199

is later than planned completion shown on the accepted programme. In short, the contractor owns the float in his programme and extensions of time are assessed with regard to the programme and not with regard to the time for completion.

## 11.2 *The compensation events*

Clause 60.1 lists eighteen compensation events which are intended to apply to all the main options.

Clauses 60.2 and 60.3 provide clarification of some of the events but do not add to the list.

Clauses 60.4, 60.5 and 60.6 are compensation events applicable only to the NEC contracts with bills of quantities – main options B and D.

The secondary options list three more compensation events:

- clause J1.2 – delay by the employer in making an advanced payment
- clause T1.1 – a change in the law occurring after the contract date
- clause U1.1 – any delay, additional or changed work caused by the application of the CDM Regulations which could not have been foreseen.

There is a space in part one of the contract data for the employer to list additional compensation events. However, it is not certain that events simply listed there become properly incorporated into the contract. There is no reference in the clauses of the contract to these additional events.

**Omissions from the list**

The most conspicuous omissions from the list of compensation events are:

- failure by the project manager or the supervisor to act in accordance with the contract
- late supply of information to the contractor
- failure by the employer to give access and use of the site to the contractor.

However, all three may be covered by other listed events.

Thus if the employer is taken to have full responsibility for the performance of the project manager and the supervisor, then any failures by them to act in accordance with the contract will be breaches of contract by the employer and covered by clause 60.1(18). Late supply of information may come within the scope of clause 60.1(3) – the employer not providing something he is required to provide by the date in the accepted programme. Failure to give access and use of the site may be covered by either clause 60.1(2) or clause 60.1(18).

**Clause 60.1(1) – changes to the works information**

Any instruction given by the project manager changing the works information is a compensation event unless it is a change:

- made to accept a defect
- made in the contractor's design at the contractor's request or to comply with works information provided by the employer.

At first sight this appears to be simply the equivalent of the usual clause in standard forms providing that all valuations shall be valued. But on closer examination it is far wider than that. It makes any change to the works information a compensation event and it is not restricted to changes in the works.

Clause 63.2 confirms that if the effect of the change is to reduce the total actual cost (e.g. an omission variation) then the contract price is reduced. This, of course, is standard practice – but see the comments under clauses 61.1 and 64.1 below on the procedural difficulties which can arise under the NEC.

Note that if the project manager permits a change in the contractor's design at the contractor's request, that is not a compensation event. So arguably if the contractor can make savings in his design he takes the full benefit of the savings (at least under Options A and B) since there is no obvious mechanism in the contract for valuing the change.

**Clause 60.1(2) – failure to give possession**

Failure by the employer to give possession of any part of the site by the later of its identified possession date and the possession date in the accepted programme, is a compensation event.

Note that the employer's obligation to give possession cannot be brought forward of the identified possession date but it can be put back. It is not wholly clear what happens if the contractor, in a revised programme, reverts to the identified possession date after showing in an earlier accepted programme a later possession date. The probability is that the employer is caught by the change and is liable under the compensation event.

Note also that clause 60.1(2) refers only to possession and not to any access as required by clause 33.2.

### Clause 60.1(3) – failure to provide something

Failure by the employer to provide something which he is contractually required to provide by the date given for providing it in the accepted programme is a compensation event.

Again there is the potential problem of dates being brought forward by the contractor in the accepted programme. And note that bringing dates forward is not a stated reason under clause 31.3 for not accepting a programme.

The expression 'does not provide something' used in the clause is potentially wide in its application and could well extend to failure to give access to the site or failure to provide information necessary for the construction of the works.

This particular compensation event is linked to dates in the accepted programme. But if, as a matter of fact, the employer is required to provide something which is necessary to the contractor's performance and there is a failure to provide by the employer, that will be a breach of contract irrespective of whether or not it is mentioned in the accepted programme. That is covered by the compensation event in clause 60.1(18).

### Clause 60.1(4) – instructions to stop or not to start work

Any instruction by the project manager to stop or not to start work is a compensation event. However, if the instruction arises from a fault of the contractor neither the contract price nor the completion date is changed (clause 61.4).

### Clause 60.1(5) – failure to work within times

Failure by the employer or others to work within the times shown on the accepted programme or within the conditions stated in the works information is a compensation event.

By this clause the employer accepts responsibility for the performance of 'others' as defined by clause 11.2(2) in so far as their performance affects the progress of the contractor. Typically, and understandably, this could include the performance of statutory undertakers and the like but the definition of 'others' is so open-ended that, in effect, the employer takes responsibility for all and sundry except those excluded by the definition.

### Clause 60.1(6) – failure to reply to a communication

Failure by the project manager or the supervisor to reply to a communication from the contractor within the time stated in the contract is a compensation event.

Under traditional forms of contract this is not normally a ground for claiming extra cost or loss and expense. Employers using in-house project managers or supervisors may be inclined to delete it from the contract, but for externally appointed project managers or supervisors it is a powerful incentive to perform to the contract.

Many have asked the question, what is the corresponding remedy for the employer if the contractor fails to reply to a communication within time? The answer is probably that in some cases the contractor prejudices his rights under the contract, but that apart the employer has no remedy unless he can prove damages and is prepared to invoke use of the disputes resolution procedures of the contract to pursue his claim.

### Clause 60.1(7) – objects of interest

An instruction given by the project manager for dealing with an object of value, historical or other interest, found within the site, is a compensation event. This compensation event arises out of clause 73.1 which is unrestricted in its application as to what constitutes an object of value, historical or other interest. The problem is likely to be what is the position if the contractor takes the clause literally and gives notice of trivialities on which the project manager refuses to give instructions.

### Clause 60.1(8) – changing a decision

A change by the project manager or the supervisor of a decision given to the contractor is a compensation event.

The intention and application of this clause is somewhat obscure. It uses the word 'decision' which is a word used very sparingly in the NEC. A 'decision' is not even listed in clause 13.1 as one of the things to be communicated in writing. The question is, does the clause apply only to the narrow range of 'decisions' mentioned in the contract or is the word 'decision' to be given wider meaning so that the clause can encompass, for example, reversals of instructions?

If the clause is given the wider application it may have the effect that any relaxation granted to the contractor by the project manager or the supervisor as a change of heart on earlier insistence of compliance in full by the contractor of his obligations is a compensation event. The acceptance of defects is dealt with under clause 44.2 so any relaxation on that front has its own mechanism for settlement; but for other issues any revocation by the project manager or the supervisor could potentially involve the employer in financial liability.

## Clause 60.1(9) – withholding an acceptance

Withholding by the project manager of an acceptance for a reason not stated in the contract is a compensation event – unless the withheld acceptance is of a quotation for acceleration or for not correcting a defect.

This clause emphasises the power of the project manager to withhold acceptance for any reason – subject to the employer being liable for any delays or costs arising unless the reason is a stated reason.

Project managers will naturally wish to avoid imposing unnecessary liabilities on the employer and accordingly they will have to be particularly careful in the wording of any non-acceptance to ensure that it conforms exactly (if that is the intention) with the wording of the chosen stated reason for non-acceptance.

## Clause 60.1(10) – searches for defects

An instruction by the supervisor for the contractor to search for a defect is a compensation event unless the search is needed because the contractor gave insufficient notice of doing work which obstructed a required test or inspection.

Because of the restricted meaning of the defined term 'defect' within the NEC, this compensation event has no application to defects for which the contractor is responsible. That apart, however, the position under clause 60.1(10) is not unlike that in traditional contracts where the contractor is liable for the cost of searches if he covers up work without giving proper notice.

## Clause 60.1(11) – tests or inspections causing delay

Any test or inspection done by the supervisor which causes unnecessary delay is a compensation event. The obligation of the supervisor to do his tests and inspections without causing unnecessary delay is stated in clause 40.5.

The main problems with this compensation event will be in distinguishing between a necessary delay and an unnecessary delay and in deciding whether the event applies only to the time taken in carrying out the tests or inspections or whether it also applies to the consequences of their findings.

## Clause 60.1(12) – physical conditions

Under clause 60.1(12) it is a compensation event if the contractor encounters physical conditions which are:

- within the site
- not weather conditions
- conditions which an experienced contractor would have judged at the contract date to have such a small chance of occurring that it would have been unreasonable to have allowed for them.

It is the rule of many contracts that employers take the risk of unforeseen ground conditions on the site. There is an element of fairness in this in that the employer has probably had more opportunity to discover the nature of the site than the contractor has had when tendering for the contract; and there is an element of commercial logic in that it may be cheaper for the employer to pay for what does happen than to pay for what might happen – assuming the contractor will price the risk if he is left to carry it.

So clause 60.1(12) is not unusual in its principles. Its wording, however, deserves careful attention.

Firstly, note that it applies when the contractor encounters 'physical conditions'. This is significant for two reasons. One, that the clause is not restricted to ground conditions; the other, that the phrase 'physical conditions' as it is applied in a similar provision in the ICE 5th edition conditions of contract has been considered by the courts and given an unexpectedly wide meaning. The Court of Appeal in the case of *Humber Oil Trustees Ltd* v. *Harbour and General Public Works (Stevin) Ltd* (1991) held that the term 'physical conditions' is not restricted to tangible matters and can also apply to a transient combination of stresses. In that case the contractor's chosen methods of working unexpectedly collapsed and the only rational explanation was that the ground had behaved unpredictably. The employer was found liable for the contractor's costs.

In short, the transfer of risk to the employer when the phrase 'physical conditions' is used may be much greater than when the phrase 'ground conditions' is used.

The second point to note is that clause 60.1(12) applies to conditions which are 'not weather conditions' and it is not restricted to conditions which are not 'due to' weather conditions as is the case in some other contracts. Consequently, weather related conditions such as floods are likely to qualify as compensation events under the NEC.

The test for the operation of clause 60.1(12) is unusually worded. It is based on what a notional experienced contractor would have judged it unreasonable to have allowed for. This is a probability test rather than an unforeseeability test and it will be interesting to see how it stands up under pressure as the NEC comes into wider use.

One aspect of the test which requires some thought is to which party would it be 'unreasonable' if the contractor does allow for physical conditions which have only a small chance of occurring. Could any fault be attributed to a contractor who does 'unreasonably' allow for such conditions and could he have any liability to the employer for doing so? The

## 11.2 The compensation events

point is raised because the question of whether the contractor has actually allowed for the physical conditions does not necessarily come into the compensation event test; and the contractor may not be debarred by the assessment rules from recovering the actual cost of dealing with the compensation event whether or not he has already included for it.

Some employers may be disposed to exclude clause 60.1(12) as a compensation event leaving the risks of physical conditions to be carried by the contractor. Some may follow the growing trend of inviting alternative tender prices with and without an unforeseen conditions clause in the contract.

There is certainly a case for excluding such clauses from contractor designed contracts on the general principle that risks should be apportioned to the party best able to control them. Where the reason for excluding unforeseen conditions clauses is to obtain better certainty of price, employers should be mindful that in excluding such clauses they increase their exposure to misrepresentation claims if the contractor can find fault with the site information provided by the employer.

### Clause 60.1(13) – weather conditions

A weather measurement, the value of which by comparison with weather data identified in the contract, is shown to occur on average less frequently than once in ten years is a compensation event if it is recorded:

- within a calendar month
- before the completion date
- at the place stated in the contract data.

This particular compensation event has not been well received by employers and it is frequently being deleted. Although it is customary to allow extensions of time for delays caused by exceptional weather it is not customary for the employer to pay the costs arising. Nor is it particularly logical that an employer should. Firstly, the employer has no control over the situation if it occurs; and secondly, the impact of an exceptional weather claim on the costs of a singular project cannot be averaged out in the same way that a contractor can average out the impact of such costs over numerous projects and numerous years.

### Clause 60.1(14) – employer's risks

If an employer's risk occurs that is a compensation event. Again, this is a controversial compensation event because it puts on the employer, amongst other things, the financial risk of certain neutral events which might be described as 'force majeure'. In traditional contracts the policy is that the risk of such events lies where it falls.

The employer's risks are detailed in clause 80.1 and they include in addition to the neutral events, claims due to faults of the employer and loss or damage due to the employer's use of the works. Many employers will carry insurance cover against the employer's risks and may be able to fund any compensation event payments to the contractor from that cover.

But that does raise a rather odd point. Under clause 83.1 the employer has to indemnify the contractor against costs due to an event which is his (the employer's) risk. However, under the compensation event procedure the employer's liability to the contractor is not for true cost but for an assessment of contractually defined actual cost – which may be more or less than true cost but is unlikely to precisely equal it. So the contract appears to give the contractor disparate remedies for the same event in the uncomfortable circumstances where a third party (the insurer) may have a valid interest in the outcome.

Employers may wish to consider with their lawyers and their insurers whether clause 60.1(14) is either necessary or desirable.

### Clause 60.1(15) – take over before completion

Certification by the project manager of any part of the works before both completion of the works and the completion date is a compensation event.

There are two important points to note here. The first is that premature use of the works (or any part) is not itself the compensation event. The trigger for the compensation event is certification of take over. The second point is that after the completion date has passed, the take over of part of the works before completion is again not the compensation event.

This latter point suggests that the employer is entitled to take over and use parts of the works after the completion date (i.e. when the contractor is in culpable delay) without incurring any liability to the contractor. Indeed at the time, the employer may well be collecting liquidated damages from the contractor for late completion.

### Clause 60.1(16) – failure to provide materials etc.

Failure by the employer to provide materials, facilities and samples for tests as stated in the works information is a compensation event. This event derives from the employer's obligations under clause 40.2.

### Clause 60.1(17) – correction of an assumption

Correction by the project manager of an assumption about the nature of a compensation event is itself a compensation event.

The reason for the inclusion in the list of compensation events of this particular event is to reconcile two other clauses of the compensation event procedure which would otherwise be in conflict:

- clause 61.6 which permits the project manager to state assumptions on which forecasts for assessments of compensation events are to be based
- clause 65.2 which states that an assessment is not revised if a forecast is shown by later information to have been wrong.

Clearly clause 61.6 could not operate fairly without the compensation event in clause 60.1(17).

Note, however, that some incompatibility still remains, in that clause 60.1(17) refers to assumptions about the 'nature' of a compensation event whereas clause 61.6 is about assumptions on the 'effects' of a compensation event.

**Clause 60.1(18) – breach of contract by the employer**

A breach of contract by the employer which is not one of the other compensation events in the contract is stated in clause 60.1(18) to be a compensation event.

This is obviously intended to be a catch-all clause for unspecified breaches of contract, and indeed it was introduced into the second edition of the NEC to cover omissions in the first edition. These led to the possibility of time being put at large by unspecified breaches of contract (since no provision existed for extending the time for completion for such breaches) and to the realisation that claims based on unspecified breach fell outside the scope of the compensation event procedure.

The problem with clause 60.1(18) is that as worded it is obviously not a true reflection of the intention of the parties. What it effectively says is that all breaches of contract by the employer are compensation events. But that is not consistent with other clauses of the contract which have their own remedies for the employer's breaches, e.g.:

- interest for late payment under clause 51.4
- payment on termination under clause 97.2.

However, this is unlikely to be fatal to the operation of clause 60.1(18) and for the contractor the clause offers a contractual remedy for breach which is potentially very wide in its scope. For the employer the clause could be seen as a general limitation of liability to assessments as made under the compensation event procedure. That could have the effect of excluding the employer's liability for consequential losses and providing the contractor with a poorer remedy than common law damages. But whether or not the clause is effective in doing that depends upon whether or not it excludes the contractor's common law rights – which is doubted.

**Clause 60.2 – judging physical conditions**

Clause 60.2 supports clause 60.1(12) by stating the factors the contractor is assumed to have taken into account in judging physical conditions. They are:

- the site information
- publicly available information referred to in the site information
- information obtainable from visual inspection of the site
- other information which an experienced contractor could reasonably be expected to obtain.

Clearly the contractor is entitled to rely heavily on the information provided by the employer and the main purpose of the clause is, perhaps, to emphasise that.

There is no requirement for the contractor to carry out his own site investigation. So if, as is often the case, tenderers are afforded the opportunity to make their own site investigations (or even required to do so) the contract requires some modifications.

**Clause 60.3 – inconsistency in site information**

Further support to clause 60.1(12) is given in clause 60.3 which states that if there is inconsistency within the site information the contractor is assumed to have taken into account the physical conditions more favourable to doing the work.

It is arguable whether or not this needs to be said since there is a rule for the legal construction of contracts (the *contra proferentem* rule) which says that in the event of ambiguities in documents the contract is construed least favourably to the party which put forward the documents.

The danger of clause 60.3 is that it appears to go much further than the *contra proferentem* rule and it arguably overrules clause 60.2. Thus, if there is an inconsistency in the site information, that inconsistency, and not the four factors listed in clause 60.2, governs what the contractor is assumed to have taken into account.

A further problem with clause 60.3, as discussed in Chapter 7, section 2, is that it appears to be wholly contrary to the contractor's duty to use reasonable skill and care when he is responsible for design.

## Clauses 60.4, 60.5 and 60.6 – compensation events for bills of quantities

For main options B and D three additional compensation events apply:

- clause 60.4 which deals with re-rating when quantities change
- clause 60.5 which deals with delays caused by increased quantities
- clause 60.6 which requires the project manager to correct mistakes in the bill of quantities.

For reasons which are obvious, bills of quantities should not be used for contracts with contractor's design unless the contract is amended to make the contractor responsible for their accuracy.

## Clause 60.4 – differences in quantities

The NEC permits tendered rates in bills of quantities to be changed if the measured quantities differ from those billed, but its rules for doing so are more elaborate than usual.

The wording of clause 60.4 is not of the clearest order and the comparison it draws between the final total quantity of work done and the quantity stated for an item is presumably to be read as the final measured quantity for an item compared to the billed quantity for the same item.

Assuming that to be the case, a compensation event occurs when:

- the difference causes the actual cost per unit quantity to change, *and*
- the measured value of the item involved is significant to the extent that it is more than 0.1% of the tender sum.

The clause expressly confirms that if the actual cost per unit is reduced, the affected rate is reduced. The intention, perhaps, is that the employer should benefit if there is an increase in quantities which reduces the cost of production. But it is not clear how this takes effect. It is unlikely that the contractor would notify a compensation event in such circumstances and the project manager has no obvious power to do so.

Where, as will often be the case, the contractor has subcontracted much of the work, the application of the clause is very much a theoretical exercise based on calculations of the defined actual cost rather than real actual cost. If this produces a windfall for the contractor that apparently is his good fortune.

## Clause 60.5 – increased quantities

A difference between the measured quantity of work for an item and the billed quantity which delays completion is a compensation event.

With this clause, as with clause 60.4, one problem is likely to be that the final total quantities which indicate whether or not there is a compensation event may not be known until after completion and, in the case of clause 60.5, any delay has been suffered.

The contractor will have to remain alert through the whole of the measurement process to the notice requirements for compensation events in clause 61.3.

**Clause 60.6 – mistakes in the bill of quantities**

The project manager is required to correct mistakes in the bill of quantities which are:

- departures from the method of measurement, *or*
- due to ambiguities or inconsistencies.

Each such correction is a compensation event.

The intention of this clause is probably no more than to follow the usual rule that the contractor prices the bill of quantities put before him when tendering and the employer accepts responsibility for any mistakes in the format of the billing. However, it may have much greater effect.

There is no exclusion in the clause for mistakes in the rates. And arguably mistakes in the contractor's rates which are due to ambiguities or inconsistencies (apparently in any of the contract documents) stand to be corrected.

Since the project manager has an obligation to correct mistakes it may be that the contractor is not required to give notice of this particular compensation event. However, the contract is not clear on the point and giving notice as a precautionary measure is, perhaps, the best course.

The reference in the clause to reduced prices is not easy to follow. It is not thought that the last sentence of the clause means that each correction may only lead to reduced prices. Nor is it too obvious how the correction of mistakes could lead to reduced prices given the definition of prices within clause 11.2(21).

## 11.3 Notifying compensation events

Clauses 61.1 to 61.7 of the NEC set out detailed requirements for the notification of compensation events.

The project manager is required to notify the contractor of compensation events arising from instructions and changed decisions; and the contractor is required to notify the project manager of other events and events not notified to him by the project manager. The latest dates for

## 11.3 Notifying compensation events

notification by the contractor are two weeks after he became aware of the event, with a final stop at the defects date.

The important question to consider is what is the significance of the notification procedure, in particular, is notification a condition precedent to the contractor's entitlement to payment? The impression has certainly been given by those promoting the NEC that the contract imposes tight time limits on the submission of claims in order to expedite their settlement and to prevent late submissions. But there are two difficulties with this. One is finding any express statement in the contract that notification on time is a condition precedent to payment; the other is the inequity of the situation (if it does exist) when applied to orders for extra works and the employer's breaches of contract.

Dealing with the second point first, clause 61.1 requires the project manager to notify the contractor of compensation events arising from instructions. If he fails to do so the obligation to notify passes to the contractor under clause 61.3. Can it really be the case, as some suggest, that the contractor loses his entitlement to payment if he fails to give his notice within the two weeks mentioned in clause 61.3? Can it possibly be intended that if the project manager orders extra works but the notification formality is delayed or omitted, the employer gets those extra works for nothing? And would the law allow such a situation when the initial breach of the notification procedure is by the employer's own project manager?

Similarly, where the compensation event is the employer's breach is it intended that the contractor's remedy is lost entirely if the contractor, the innocent party, fails to give notice within two weeks?

Neither common sense nor equity allow either of these, which leads to the probability that either the notification procedures are strictly binding – but the compensation event procedures are optional remedies and not exclusive remedies – or the notification procedures are not strictly binding such that they disqualify late submissions under the procedure.

Turning to the point on the absence of any express statement on the consequences of failure to comply with the notice requirements, the nearest the contract comes to imposing the compensation event procedure as mandatory is that the various definitions of the prices for each of the main options include the phrase 'unless later changed in accordance with this contract'. The nearest the notification clauses come to suggesting that they act as conditions precedent to payment is that clause 61.3 puts an obligation (although perhaps it is meant as an entitlement) on the contractor to notify a compensation event if 'it is less than two weeks since he became aware of the event'. On one reading of the clause this suggests the contractor is not obliged to notify an event if it is more than two weeks since he became aware of it.

The problem of interpretation here is that the phrase 'the Project Manager notifies' in clause 61.1 is almost certainly a statement of obligation, whereas the phrase 'the Contractor notifies' in clause 61.3 may not be meant as a statement of obligation but as a statement of entitlement – although this latter meaning would be a departure from the general rule of style of the NEC that a present tense verb implies the word 'shall'.

But whatever the meaning or intention of clause 61.3 it cannot be said that it expressly prohibits the notification of compensation events at whatever time the contractor chooses. And significantly clause 61.4 which lists the circumstances in which the contract price and completion date are not changed does not include in its list an item for late notification of an event.

**Clause 61.1 – notifications by the project manager**

Clause 61.1 refers only to compensation events arising from:

- instructions given by the project manager or the supervisor
- changes of earlier decisions given by the project manager or the supervisor.

Most obviously this would apply to compensation events 1, 4, 7, 8, 9, 10, 15 and 17, but this list may not be conclusive.

For such events the project manager is required to notify the contractor of the compensation event 'at the time of the event'. He is also required to instruct the contractor to submit quotations unless:

- the event arises from a fault of the contractor, *or*
- quotations have already been submitted.

The administrative procedures involved in this are quite formidable. Clause 13.7 requires any notification to be communicated separately from other communications – so each time the project manager or the supervisor gives an instruction or change of decision within the scope of clause 61.1, a separate notice should be sent, simultaneously but separately, giving notice that the instruction or changed decision is a compensation event.

In the event of the project manager failing to give notice, the burden passes to the contractor under clause 61.3, with the implications discussed above. The failure may itself also be a compensation event under clause 60.1(18) – breach by the employer (on the basis that the employer is responsible for the project manager's performance).

An interesting contractual situation arises when the project manager notifying a compensation event does not instruct the contractor to submit a quotation, in the belief that the event arises from a fault of the contractor. Clauses 61.3 and 61.4 are by-passed because the event has already been notified; and clause 62.3 on the requirement to submit a quotation does not apply because there has been no instruction to do so. However, if the contractor, of his own accord, chooses to submit a quotation under clause 62.2, the project manager's authority appears to be limited to deciding whether the assessment has been correctly made.

Eventually, no doubt, any dispute on the matter would reach the adjudication provisions of clause 90.1, but much then depends on how the dispute is framed as to what follows next – particularly in the period before settlement – because on the wording of clause 90.2 the project manager is required to proceed as if the disputed matter was not disputed.

The last sentence of clause 61.1 requiring the contractor to put an instruction or changed decision into effect appears superfluous in the light of clause 29.1 which requires the same.

### Clause 61.2 – quotations for proposed instructions

Clause 61.2 permits the project manager to instruct the contractor to provide quotations for proposed instructions or proposed changed decisions.

This is a sensible provision spoilt somewhat by the definition in clause 11.2(28) applying to main options A and B, indicating that the cost of preparing quotations for compensation events is excluded from actual cost. It cannot be right that the contractor should be put to the expense of preparing quotations without recompense. Perhaps the contractor can argue that the instruction to provide a quotation is itself a compensation event.

An oddity in the wording of the clause is that the quotations are for a proposed instruction or a proposed decision and not for the compensation events flowing from them.

### Clause 61.3 – notification by the contractor

Clause 61.3 states that the contractor 'notifies' an event which has happened or which he expects to happen as a compensation event if:

- the contractor believes it is a compensation event
- it is less than two weeks since 'he' (presumably the contractor) became aware of the event
- the project manager has not notified the event to the contractor.

Comment on the meaning of 'notifies' and the general application of this clause has been given above.

Note that the question of 'fault' is not relevant to the application of clause 61.3. That follows in clause 61.4.

On the wording of the proviso 'it is less than two weeks since he became aware of the event', note that it is expressed as a factual test 'he became aware of' and not as an objective test 'he should have become aware of'. This is likely to be a source of much relief to contractors.

### Clause 61.4 – the project manager's decisions

The opening words of clause 61.4 convey its importance: 'The Prices and the Completion Date are not changed if...'. The clause then goes on to set out the circumstances in which there will be no change. Note, however, that the clause applies only to decisions by the project manager on events notified by the contractor.

The decisions which preclude change are that an event:

- arises from a fault of the contractor
- has not happened and is not expected to happen
- has no effect on actual cost or completion
- is not one of the compensation events in the contract

If the project manager reaches any other decision the clause requires that he should then instruct the contractor to submit quotations for the event.

The project manager is required to notify the contractor of the decision he has reached on a compensation event:

- within one week of the contractor's notification, *or*
- within such longer period as the contractor agrees.

For complex issues the project manager may find the one week allowed to him for making his decision less than adequate but he is reliant on the contractor's agreement to any extension of the period. Whether or not that agreement is forthcoming may well depend upon how the contractor interprets the contract on what happens if the project manager fails to give his decision in time. The contract is not clear on this but one possibility is that if the contractor proceeds to submit a quotation of his own accord, the project manager is stopped from belatedly ruling it out under clause 61.4.

Users of the first print of the second edition of the NEC should note that in the second print (date stamped November 1995) a change in the position of a full stop in clause 61.4 significantly altered its meaning.

### Clause 61.5 – failure to give early warning

The early warning scheme of clause 16 is intended to be more than simply persuasive and failure by the contractor to comply with its notice provisions is taken into account in assessing compensation events.

Clause 61.5 commences the procedure for this by requiring the project manager to notify the contractor if he decides that the contractor did not give an early warning which an experienced contractor could have given. The project manager is to notify his decision to the contractor when he instructs him to submit quotations.

The purpose of the latter provision is perhaps to indicate that the project manager has only one opportunity to raise alleged lack of early warning and that he cannot notify the contractor after he has instructed him to submit a quotation.

### Clause 61.6 – assumptions on effects

Bearing in mind the speed at which the notification and quotation procedures for compensation events are intended to operate, it is essential that some provision is made for uncertainty of effects. Clause 61.6 does this, but not perhaps as generously as contractors would prefer, because only the project manager is allowed to make assumptions on the effects of compensation events which are correctable.

The clause provides that:

- if the project manager decides that the effects of a compensation event are too uncertain to be forecast reasonably
- he states assumptions about the event in his instruction to the contractor to submit quotations
- the assessment of the event is based on these assumptions
- if any assumption is found to be wrong the project manager notifies a correction (and compensation event 60.1(17) then applies).

The full implications of this clause can only be seen by reference to clause 65.2 which states that the assessment of a compensation event is not revised if a forecast is later shown to have been wrong. The question is who is to take the risk on inaccurate forecasts – the contractor or the employer?

If the contractor perceives that he is at risk by the uncertainty of effect of a compensation event he will load his quotation so as to minimise that risk. The project manager may then see the employer as being at risk of paying over the odds and may see himself in the firing line for having permitted it.

Some project managers using the NEC have confirmed a policy of automatically stating assumptions. They do this as a precautionary measure to retain some element of cost control. Their concern is that the NEC provisions for the non-acceptance of quotations and assessments for compensation events may not extend to challenging the contractor's assumptions and may be limited to challenging only his calculations.

This may well be a correct analysis of the contract – see the comment in section 6 below on clause 64.1.

### Clause 61.7 – no notification after the defects date

Clause 61.7 states briefly that a compensation event is not notified after the defects date.

Provisions of similar effect are found in most standard forms with the purpose of finalising the date by which the contractor can submit contractual claims. Most, however, allow three months or so after the end of the defects correction period, whereas the NEC specifies the defects date as the cut-off – which is the end of the period.

In some circumstances the NEC cut-off could conflict with other clauses of the contract. Note that under clause 82.1 the contractor's obligations extend until the defects certificate is issued and that under clause 43.2 the issue of the defects certificate may be later than the defects date.

## 11.4 Quotations for compensation events

Quotation is not a defined term of the NEC but from clause 62.2 it can be seen to be a proposal from the contractor for changing the contract price or extending the contract time.

### Clause 62.1 – instructions for alternative quotations

Clause 62.1 permits the project manager to instruct the contractor to submit alternative quotations based on different ways of dealing with compensation events. It requires the contractor to submit quotations as so instructed and permits him to submit quotations for other methods which he considers practicable.

Although the clause refers to 'different ways' and the 'other methods', which suggests that it is intended to deal principally with practical alternatives, it may be wide enough to permit different combinations of price and time for the same method. Used in this way it would act as a supplement to clause 36.1 which provides for acceleration in that the employer could effectively purchase the contractor's right to more time by agreement.

As with clause 61.2 the expense of submitting the quotations is left with the contractor under main options A and B.

On its wording, clause 62.1 looks as though it can only be activated by

## 11.4 Quotations for compensation events

the project manager but it is possible that taken together with the provisions in clause 16.3 for proposals at early warning meetings, the contractor may be able to activate it of his own accord.

### Clause 62.2 – quotations for compensation events

Clause 62.2 starts with the important definition that quotations for compensation events comprise proposed changes to the prices and any delay to the completion date assessed by the contractor. Taken by itself that would suggest that it is a delay beyond the completion date which has to be assessed. But note clause 63.3 which clarifies what is meant in clause 62.2 and makes clear that it is a delay to completion which is to be assessed.

Clause 62.2 continues with the requirement that the contractor is to submit details of his assessment with each quotation and if the programme for the remaining work is affected by the compensation event the contractor is to include a revised programme in his quotation.

The amount of detail required depends not only on the scale of the compensation event involved but also on which of the schedules of cost components is used for assessment – see clause 63.11. However, whichever schedule is used the accepted programme (or its revision) is the key to the assessment since the actual cost basis for assessment relies on cost components and duration times.

### Clause 62.3 – submission of quotations

The contractor is required to submit quotations within three weeks of being instructed to do so. The project manager is required to reply within two weeks of the submission. These are tight times but note the possibility of agreement of an extension under clause 62.5.

Failure by the contractor to submit his quotations on time entitles the project manager to make his own assessments (clause 64.1), but the consequences of failure by the project manager to reply in time are less clear. Certainly, there is the possibility of a new compensation event arising under clause 60.1(6). But no doubt what the contractor would like to know is whether failure to reply in time has the effect of preventing any belated attempt by the project manager to state non-acceptance. Given the rigid administrative rules of the NEC that may well be the case.

Clause 62.3 emphasises the point on rules because it does not leave it to the project manager to decide what type of reply he should give. The clause restricts his reply to four forms:

- an instruction to submit a revised quotation
- acceptance of the quotation

- notification that a proposed instruction or changed decision will not be given
- notification that he will be making his own assessment.

Of these the least certain in its intention and application is the first one – an instruction to submit a revised quotation. Clause 62.4 requires the project manager to explain his reasons for asking for a revised quotation but it says nothing as to what those reasons might be. The question is, can the project manager question the logic of the contractor's quotation? Can he substitute his own assumptions on effects for the contractor's assumptions? Some thoughts on this have been given under clause 61.6 above.

Certainly the project manager is permitted to question the mechanics of the contractor's calculations – that is allowed by clause 64.1 – and note that in that clause the reference to 'a revised quotation' relates to a quotation which has not been assessed correctly.

### Clause 62.4 – instruction to submit a revised quotation

Clause 62.4 permits the project manager to instruct the contractor to submit a revised quotation – but only after explaining his reasons for doing so.

The contractor is required to submit his revised quotation within three weeks (as clause 62.3). But clause 62.5, allowing for extended time by agreement, appears to apply to revised quotations as well as to original quotations.

One aspect of this clause which is not too clear is whether it can be used on a repeat basis. It probably can if only because the consequences of not being able to do so would be too advantageous to the contractor.

### Clause 62.5 – extending the time for quotations

The project manager is permitted to extend the time allowed for the contractor to submit quotations and for his own replies if he and the contractor agree to the extension before the submission or reply is due. The project manager is required to formally notify the contractor of any extension agreed.

This provision for relaxation of the timescales will be welcomed by both the contractor and the project manager as a useful change from the rigidity of the first edition of the NEC, but note that it operates only by agreement.

## 11.5 Assessing compensation events

The rules set out in common core clauses 63.1 to 63.7 and main option core clauses 63.8 to 63.11 for the assessment of compensation events will tax the brains of quantity surveyors for many years to come. They certainly deserve, and will probably get in the course of time, a comprehensive analysis of their complexities. All that can be done in this book is to indicate the apparent intention of the rules and some of the more obvious problems in their application.

**Clause 63.1 – changes to prices**

Clause 63.1 states that the changes to the prices are assessed as the effect of the compensation event upon:

- the actual cost of the work already done
- the forecast actual cost of the work not yet done, *and*
- the resulting fee.

The broad intention of this is clearly that all claims and variations are to be valued on a cost basis. The traditional use of bill of quantity rates and prices for valuing variations is discarded.

In principle this should be fair to both parties – or at least so for variations. For claims it may not be so fair to the contractor since claims valued on cost may not provide as good a remedy as claims valued as damages. In practice the general difficulty with cost based valuations for claims and variations is that usually it is extra cost which is to be valued. And identifying extra cost either in extent or in records is no easy matter.

Clause 63.1 of the NEC suffers, as it must, from the general difficulties of cost based valuations but it greatly adds to those difficulties by using a cost base, which although called actual cost is, for the two important main options A and B, a long way from real actual cost. So firstly there is a records problem, and secondly there are the complexities of notional assessments. Another aspect of concern is that there is no requirement that cost should be incurred; it is the notional calculation of assessed actual cost which is considered, not the reality of actual incurred cost. With main options A and B, the consequences of this if much of the work is subcontracted can only be described as unpredictable.

For example, the first part of clause 63.1 states: 'The changes to the Prices are assessed as the effect of the compensation event upon the Actual Cost of the work already done.' This part applies to compensation events which are being valued retrospectively.

It requires that a comparison should be made between the value of the work completed as if the compensation event had not occurred and the value of the work completed including the effects of the compensation event – in both cases using the actual cost formula of the contract to provide a common base for comparison. Now the one thing that any contractor using main options A or B is unlikely to have in his records is the value of the work in accordance with the actual cost formula. What he will have are records of real costs – and for the most part they are likely to be subcontractor's accounts – a category of cost not recognised in the definition of actual cost for main options A and B. He will also know the value of the work completed in accordance with his activity schedule or his bill of quantities. These are of no help to him. What he has to do is a notional calculation of what the work he has completed would have cost, with and without the compensation event, using the contractual actual cost formula.

Applying the same principles to omission variations is even more tortuous and the result even more unpredictable.

It has been suggested by some commentators on the NEC that this is such an elaborate and potentially costly scheme that there must be some other interpretation of the first part of clause 63.1 to produce a simpler result. That would certainly be so if it was possible to take the phrase 'the work' as applying to the work involved in the compensation event. Unfortunately that is not possible given the reference to 'the effect of the compensation event upon ... the work'.

Another attempt at simplification has suggested that actual cost in clause 63.1 is not to be taken as following the definitions of actual cost in respect of main options A and B in that the reference in the definition to 'the components in the Schedule of Cost Components' should be excluded. This idea has much to recommend it in that it would allow the cost of work already done to be valued simply by reference to records or to be taken from the activity schedule or the bill of quantities. Again, unfortunately, the wording of the contract does not support this approach.

The second part of clause 63.1 deals with forecasts of actual costs. Again the basis is the contractual actual cost and the comparison between forecasts with and without the compensation event. Costs as indicated in an activity schedule or bill of quantities have no relevance. Again, for main options A and B subcontractor involvement may produce some complexities and some highly artificial assessments.

But, of course, as with all forecasts the really contentious area is likely to be in the validity of the assumptions on which forecast is based. The NEC does try to regulate this by relating the forecast to an accepted programme so that the resources involved in the forecast are quantified and not simply guessed.

It is not clear whether it is permissible in either retrospective or forecast assessments under clause 63.1 to group compensation events together into a single assessment. It will probably be sensible to do so in some cases and given the notional element of the calculations it is not obvious why there should be any objection.

## 11.5 Assessing compensation events

The reference in clause 63.1 to the fee is to ensure that assessments can include the fee as defined in clause 11.2(17). Accordingly the fee is the fee percentage as stated by the contractor in part two of the contract data applied to the amount of actual cost.

An interesting point is to consider how this applies to omission variations. Is the fee in such a case a positive or a negative amount? Clause 63.2, which deals with reductions in actual cost and in the contract price, does not mention the fee. On arithmetical logic a negative actual cost should produce a negative fee but there can obviously be injustice in that to the contractor.

A greater potential for injustice, however, with regard to the fee is in connection with subcontractors. The contractor's fee is deemed to include for a subcontractor's fee. There will be no injustice providing contractors are permitted to quote a higher percentage fee than their subcontractors. But some employers are known to baulk at contractors' fees in excess of 5%, notwithstanding the fact that specialist subcontractor fees can be as high as 25%.

### Clause 63.2 – reductions in prices

Clause 63.2 restricts reductions in the contract price to reductions arising from certain types of compensation events. These are:

- a change in the works information (e.g. an omission variation)
- a correction of an assumption made by the project manager in assessing an earlier compensation event (i.e. clause 60.1(17) compensation events) which can lead to reductions.

Note also, however, that clauses 60.4 and 60.6 of main options B and D relating to remeasurement allow for reductions in the contract price. So also does secondary option T1 on changes in the law.

### Clause 63.3 – delay to completion

Clause 63.3 states that any delay to the completion date is assessed as the length of time that due to the compensation event planned completion is later than planned completion shown on the accepted programme.

For contractors this is a particularly important clause. It indicates that the contractor's entitlement to an extension of time for completion is judged by reference to the date of planned completion on his accepted programme. Any assessed delay beyond that date caused by a compensation event is added to the formal contract time for completion by adjusting the stated completion date. Thus if the contractor has float in his programme he retains that float.

It is slightly misleading that the clause starts with the phrase 'A delay to the Completion Date' but that is clarified later in the clause. Perhaps it would also have been helpful if the clause had clarified its closing words 'Accepted Programme'. Presumably that is meant to be an accepted programme prior to the effect of the actual or forecast effect of the compensation event being recorded. But given that the term 'accepted programme' has a defined meaning in clause 11.2(14), that may fix the accepted programme applicable to clause 63.3 – with potentially the wrong results if a revised programme with the effects of the compensation event has become the accepted programme by the time of the assessment.

### Clause 63.4 – failure to give early warning

Clause 63.4 contains the sanction on the contractor for failing to give an early warning of a compensation event. The clause states that:

- if the project manager has notified the contractor of his decision
- that the contractor did not give an early warning
- which an experienced contractor could have given
- the event is assessed as if the contractor had given early warning

At first reading it does appear that if an event is assessed 'as if the Contractor had given early warning' it does not matter whether the contractor gave early warning or not. But obviously the intention of the clause must be the opposite.

In practice, however, applying the clause with this latter intention could lead to some surprises. It may be the expectation that the contractor will suffer because of his default of not giving early notice – but it does not follow that this will be the result, particularly having regard to clause 65.2.

### Clause 63.5 – time and risk allowances

Clause 63.5 can be seen as supporting clauses 63.1 and 63.3 which require the 'effects' of a compensation event to be assessed. It expressly permits assessments to include for risks of cost and time. The clause states:

- the assessment of the effect of a compensation event
- includes cost and time risk allowances
- for matters which have a significant chance of occurring
- and are at the contractor's risk.

Some matters such as materials wastage, downtime due to inclement

## 11.5 Assessing compensation events

weather and other routine estimating allowances should be relatively non-controversial. More sophisticated allowances such as provision for excesses on damage claims, may attract closer examination and argument. The big issues, however, will be delay, disruption and winter working. It is up to the contractor to show that they have a 'significant' chance of occurring.

The intention of the words which conclude clause 63.5 – 'and are at the Contractor's risk under this contract' – is not immediately obvious. The clause would seem to work perfectly well without the words. When applied to the main options with a cost reimbursable element they seem to throw some uncertainty on whether risks should be allowed for in whole or in part.

### Clause 63.6 – assumptions on reactions

Clause 63.6 states certain restrictive assumptions applied to assessments:

- that the contractor reacts competently and promptly
- that additional cost and time are reasonably incurred
- that the accepted programme can be changed.

The first two of these are common sense measures designed to protect the employer against the contractor's inefficiencies. They correspond broadly to the rules for the assessment of damages. The third point in the clause, the reference to the accepted programme, is an indication that the contractor is expected to change his programme if that is practicable, but that begs the question, what is the position if the accepted programme cannot be changed? Consider, for example, a programme which is fixed by external restraints such as railway track possessions. Is the contractor's entitlement in such circumstances to be paid the actual costs of working to the fixed programme or is he entitled to the notional costs of working to a changed programme? The latter would have some logic in that the contractor would be paid the costs of a notional extension of time instead of being paid his acceleration costs.

### Clause 63.7 – ambiguity or inconsistency

Clause 17.1 requires the project manager to give an instruction resolving any ambiguity in or inconsistency between the documents in the contract. This is a general requirement which covers both employer and contractor provided documents. The probability is that any instruction given will change the works information but the possibility is that the fault will not always be the employer's.

Clause 63.7 deals with this by stating how compensation events which are instructions to change the works information to resolve an ambiguity or inconsistency, are to be assessed. It follows a legal rule for the construction of contracts (the *contra proferentem* rule) in allowing the most favourable interpretation of the documents to be given to the party which did not create the ambiguity or inconsistency.

So the clause provides that if the employer produced the works information which is at fault, the compensation event is assessed with an interpretation in favour of the contractor. If the contractor produced the works information, the effect of the compensation is assessed with an interpretation in favour of the employer.

It does appear that the clause goes a little further than traditional contractual provisions to the same effect in that, taken with clause 63.2 which allows for reductions in the contract price for changes in the works information, it may give the employer rights to what are, in effect, contractual rights of counterclaim, for faults in works information provided by the contractor. Alternatively, or additionally, it permits the project manager to instruct the contractor to perform the works to the highest standards that his (the contractor's) documents suggest, without the employer taking on liability for any extra costs.

### Clause 63.8 – changes to the activity schedule

Clause 63.8 applies only to main options A and C. It states that assessments for changed prices for compensation events are in the form of changes to the activity schedule.

This is a curious and unusual arrangement but one which is fully in keeping with the approach of the NEC to lump sum contracts in that the contract is structured to recognise the components of the lump sum as individual contract prices. Amounts due on claims and variations are not added directly to the contract price but are first added to items in the activity schedule. Presumably in many cases this will require apportionment of the assessment, the sole purpose of which would seem to be to maintain the integrity of the interim payment scheme.

### Clause 63.9 – changes to the bill of quantities

Clause 63.9 applies only to main options B and D. It has two provisions. Firstly, the clause states that changed prices for compensation events are in the form of changes to the bill of quantities. Secondly, the clause states:

- if the project manager and the contractor agree
- rates and lump sum prices in the bill of quantities

- may be used as the basis for assessment
- instead of actual cost and the resulting fee.

The first provision may not be as onerous as it at first appears. Presumably instead of allocating an amount due on a compensation event to various rates and prices, it is permissible to include it in the bill of quantities as a lump sum.

The second provision recognises the difficulties of notional actual cost calculations and allows an easier method of assessment.

### Clause 63.10 – subcontracted work

Clause 63.10 applies only to main options A and B. It develops the note which heads the schedules of cost components and indicates that for options A and B references to the contractor in the schedules include for subcontractors. In other words, for options A and B, subcontract invoices are not accepted as actual costs.

The clause confirms that only the contractor's fee percentage is added to actual cost, even if the work involved is subcontracted. Any fees paid by the contractor to the subcontractor are not included in the assessment.

Ideally then the contractor, for his own protection, should always quote a fee percentage which is higher than that of his subcontractor's level of on-cost.

### Clause 63.11 – use of the shorter schedule

Clause 63.11 applies to main options A, B, C, D and E. It has no relevance to main option F.

The clause has two provisions:

- the contractor can assess a compensation event using the shorter schedule of cost components if the project manager agrees
- the project manager is entitled to make his own assessments using the shorter schedule.

For comment on the differences between the schedules of cost components see Chapter 4, section 9.

Briefly, the shorter schedule is simpler and speedier but it puts more risk on the contractor.

## 11.6 The project manager's assessments

Clauses 64.1 to 64.3 set out the circumstances in which the project manager may make his own assessments of compensation events.

By doing so they reveal an implication in the NEC that (the stated circumstances apart) the contractor is entitled to value claims and variations. This, if it is the case, is significantly different from the position in traditional standard forms. It raises some interesting questions on how disputes should be progressed. In particular, should the contractor's assessment stand as valid unless and until changed by the adjudicator; and is there even provision for challenging the contractor's assessment outside the provisions of clauses 64.1 and 64.2?

### Clause 64.1 – the project manager's assessment

Clause 64.1 requires the project manager to assess a compensation event:

- if the contractor has not submitted his quotation and assessment in time
- if the project manager decides that the contractor has not assessed the event correctly (and no instruction to submit a revised quotation has been given)
- if the contractor has not submitted with his quotation a required programme
- if when the quotation is submitted the project manager has not accepted the contractor's latest programme for a reason stated in the contract.

The broad intention of this clause appears to be that the project manager can, and must, intervene in the assessment process if the rules on which it relies are not being followed.

However, it is certainly arguable that there is nothing in the clause permitting the project manager to make his own assessment simply because he disagrees with the assumptions made by the contractor in his assessment. The clause appears to be concerned with the mechanics of assessment and not with the logic of assumptions.

For comment on the situation if the contractor disagrees with the project manager's assessment see Chapter 14, section 4.

### Clause 64.2 – assessed programme

Clause 64.2 is a long stop provision which permits the project manager to make his own assessment of the programme for the remaining work when:

- there is no accepted programme
- the contractor has not submitted a revised programme for acceptance as required.

## 11.6 The project manager's assessments

This should act as a powerful incentive to the contractor to ensure that an accepted programme is rapidly in place and that it is properly revised.

**Clause 64.3 – notification of the project manager's assessments**

When the project manager makes an assessment of a compensation event he is required to notify the contractor and to give details of the assessment within the same period allowed to the contractor for submission of a quotation (three weeks under clause 62.3). The period starts when the need for the project manager's assessment becomes apparent.

The consequences if the project manager fails to comply with the timescale are not stated but it is possible that the effect is to invalidate his assessments.

## 11.7 Implementing compensation events

The NEC expresses the process of changing the contract price or extending the time for completion as 'implementing compensation events'. The project manager is required to undertake this task.

The phraseology is curious since by clause 60.1 compensation events are stated to be various actions, circumstances or breaches and in the ordinary meaning of words they are not capable of being implemented by the project manager.

**Clause 65.1 – implementing compensation events**

Both the wording and the structure of clause 65.1 are odd.

The first provision requires the project manager to implement 'each compensation event' by notifying the contractor of the quotation he has accepted or of his own assessment. The second provision states that the project manager implements the compensation event at the later of:

- when he accepts a quotation, *or*
- when he completes his own assessment, *or*
- when the compensation event occurs.

It is not absolutely clear how the two provisions fit together but the intention is perhaps that implementation is a two stage process, the first stage being notification of intended changes to the contract price and time, and the second stage being the application of the changes.

This interpretation is supported by the reference in the second provision to 'when the compensation event occurs'. At first sight it is difficult to see how this can be later than either the quotation or the assessment, but presumably it applies in respect of future events in the sense that the changed prices and time do not take effect at the notification but are only 'implemented' later when the event actually occurs.

### Clause 65.2 – no revision for later information

The implications of clause 65.2 have been discussed in section 11.3 above, under clause 61.6.

Clause 65.2 is brief but very important. It states that the assessment of a compensation event is not revised if a 'forecast' upon which it is based is shown by later 'recorded' information to have been wrong.

In short an assessment which is accepted stands as valid even if the assumptions in the forecast on which the assessment is based are later proved to be wrong. Or put another way, the contractor stands the risks of his forecasts. He may lose out financially or make a surprise windfall. But what has been achieved is a measure of certainty on the amount of money to change hands and the amount of time to be awarded.

Note that by using the word 'forecast' in relation to what may or may not have been wrong, the clause leaves open the possibility of a revision of an 'assessment' which is shown to be wrong.

One aspect of the clause which is certain to attract debate is whether it confers absolute finality on forecasts. And, if it does, whether this applies only to forecasts made by the contractor or also to those made by the project manager.

The significance of the word 'recorded' in the last line of the clause is not fully understood. The clause confirms a negative position, so qualifying the information by 'recorded' could suggest limitation and not enhancement of the application of the clause. More likely, what it means is that no account is to be taken in any re-assessment, of information which was not available at the time of the assessment.

### Clause 65.3 – changes to forecast amounts

Clause 65.3 applies only to main options E and F – the two fully cost reimbursable options. It supports clause 65.1.

The clause states that the project manager includes the changes to the forecast amounts of the prices and the completion date in his notification to the contractor implementing a compensation event. This appears to be a reference to the forecast which the contractor is obliged to prepare under clause 20.4, but whatever its meaning it highlights a fundamental question – what purpose does the compensation event procedure serve in fully cost reimbursable contracts?

### Clause 65.4 – changes to prices and the completion date

Clause 65.4 applies to main options A, B, C and D. It states that the project manager includes the changes to the prices and the completion date which he has accepted or assessed in his notification implementing a compensation event.

Again, this is no more than detail supporting clause 65.1.

### Clause 65.5 – subcontract compensation events

Clause 65.5 applies only to main options E and F. It states that the contractor does not implement a subcontract compensation event until it has been agreed by the project manager. The purpose is to give the project manager control over subcontract costs before they are incurred and become due for reimbursement.

# Chapter 12

# Title

## 12.1 Introduction

Section 7 of the NEC deals with title to equipment, plant and materials, and other objects of value or interest.

The NEC avoids the ambitious provisions of many standard forms of contract which purport to give the employer title to anything he has paid for or anything the contractor has brought onto site. Instead the NEC sensibly goes no further than passing to the employer 'whatever title the Contractor has'.

This may alarm some employers who want to be assured that payment for plant and materials secures ownership, and that equipment brought onto site for the construction of the works (e.g. falsework and scaffolding) cannot be removed until it has served its purpose. The concern of such employers is understandable but it is better addressed by the realistic provisions of the NEC and firm discipline by the project manager and the supervisor in checking the contractor's title, than by the misleading and ineffective assertions to title of some contracts.

The problem with such contractual assertions, which the NEC recognises, is that they operate only between the employer and the contractor and they do not diminish the legal rights to title of third parties or override statutory rules in the event of insolvency. They may give comfort to the employer by their appearance but when things go wrong and conflicting claims to title emerge, then the full complexity of the law of ownership is revealed.

It is beyond the scope of this book to comment in detail on that law except to say that amongst the complexity one firm rule has remained clear and withstood the passage of time. That is the rule in *Appleby* v. *Myers* (1867) which states that 'Materials worked by one into the property of another become part of that property'. In short, title passes to the employer as against the contractor when plant and materials are incorporated into the works. Anything more complicated than that is a job for the lawyers.

## 12.2 Employer's title to equipment, plant and materials

### Clause 70.1 – title outside the working areas

Clause 70.1 provides that whatever title the contractor has to equipment, plant and materials:

- which is outside the working areas, *and*
- has been marked by the supervisor as for the contract, *then*
- that title passes to the employer.

The purpose of this clause is quite modest – simply to state the change of title effected by marking. The clause says nothing about the obligation or the entitlement to mark (for that see clause 71.1) nor anything about the contractor's rights (if any) to payment consequent upon marking – for comment on that see section 12.3 below.

Note, that the clause refers to the 'working areas' and not to site. This maintains compatibility with the schedules of cost components.

### Clause 70.2 – title within the working areas

Clause 70.2 deals with equipment, plant and materials brought within the working areas. It provides that whatever title the contractor has passes to the employer.

The clause further provides that title passes back to the contractor when equipment, plant and materials are removed from the working areas with the project manager's permission.

The transfer of title under this clause is automatic and is not dependent on marking by the supervisor. Nor is it expressly made dependent on payment or any right to payment. This raises interesting questions on whether the transfer of title, particularly with respect to the contractor's equipment, is intended to be permanent or temporary. For example, if the contractor becomes insolvent, is the employer entitled only to retain title to the equipment until completion of the works, or can the employer sell the equipment after completion and retain the proceeds notwithstanding any claim on the equipment by a receiver or liquidator?

The answer to this may be found in clause 96.2 (termination) which states that the employer 'may use' any equipment to which he has title. This suggests that the transfer of title, at least for equipment, is intended to be only temporary.

In so far as parts of the working areas are outside the boundaries of the site, the employer's hold on items to which he acquires title may be somewhat precarious. The employer can secure his site to protect his hold on items to which he claims title, but the working areas outside the site may be beyond his control. This may present a problem for those employers who, as a matter of policy, decline to pay for any goods or materials until they are brought within the boundaries of the site or the employer's premises.

Note that the reversal of title to the contractor under clause 70.2 only occurs when removal of equipment, plant and materials from the working areas is with the project manager's permission. There is no express obligation in the NEC to seek permission for removal and no obvious prohibition on removal without permission. However, the potential problem for the contractor is that if he does not seek permission for removal, title remains with the employer and the consequences of this could be unpredictable.

The consequences of the project manager refusing to give permission for removal are also unpredictable. This situation is not expressly covered in the contract as a compensation event and a claim by the contractor for breach of contract would be difficult to sustain.

## 12.3 Marking equipment, plant and materials

Clause 71.1 supports clause 70.1 in detailing the circumstances in which equipment, plant and materials outside the working areas are to be marked.

**Clause 71.1 – marking equipment**

Clause 71.1 provides that the supervisor marks such equipment, plant and materials if:

- the contract identifies them for payment, *and*
- the contractor has prepared them for marking as the works information requires.

The dual aspects of this clause cause some difficulties of interpretation. The intention of the clause, taken together with clause 70.1, appears to be that if the employer wants to obtain title to equipment, plant or materials before they are brought within the working areas, then he must pay for the privilege and in return the contractor must allow the relevant items to be marked as the property of the employer by the supervisor. Or alternatively, if the contractor wants to obtain the benefit of payment for off-site equipment, plant and materials, then he must allow them to be marked and concede the transfer of title.

What is not clear is where, and by which party, the identification for payment is to be made and whether the requirements in the works information are intended to be procedural or item specific. What, for example, is the position if the contractor in his activity schedule or bill of quantities identifies off-site items for payment but the works information is silent on marking? Is that to be taken as an indication that the employer does not intend to pay for off-site items, or as merely an administrative omission in the works information which can be corrected by the project manager?

These are points the parties would do well to clarify before they sign their contract.

## 12.4 Removing equipment

Construction contracts traditionally include clauses requiring the contractor to clear the site on completion and to remove items belonging to the contractor. Such clauses may not be strictly necessary since the obligations can probably be implied as integral and essential parts of proper performance.

The NEC does not have the usual clearance of site on completion clause, but in clause 72.1 it does deal with removal of the contractor's equipment.

### Clause 72.1 – removal of equipment

Clause 72.1 requires the contractor to remove equipment from the site when it is no longer needed, unless the project manager allows it to be left in the works. Note that in this clause the obligation is expressed, quite rightly, as to remove from the site and not from the working areas.

The obligation is stricter than that found in most contracts in that it may take effect before completion. However, if the contractor is in breach it would take exceptional circumstances for the employer to have a legal remedy – and there is nothing in the contract to suggest that the employer has the right to effect the removal himself.

One example of exceptional circumstances could be where contractor's equipment which is no longer needed for the works is obstructing the employer's use of his premises. Perhaps then the employer could exercise his right to terminate under clause 95.3 (reason R14).

## 12.5 Objects and materials within the site

### Clause 73.1 – articles of interest

Clause 73.1 contains four distinct provisions:

## 12.5 Objects and materials within the site

- the contractor has no title to an object of value or historical or other interest found on the site
- the contractor is required to notify the project manager when such an object is found
- the project manager is required to instruct the contractor how to deal with the object
- the contractor is not permitted to move the object without instruction.

This is the equivalent of a typical 'antiquities' clause and in principle it is perfectly sound. Unfortunately the wording of the clause exhibits a tendency running through the NEC of imposing unqualified obligations without regard to practical applications. The problem is, who is to decide whether a thing is of value or interest – the project manager or the contractor? If it is the project manager and the contractor takes clause 73.1 literally, work will stop every time a coin, bone or fossil is found. Obviously the clause must be given some common sense level of application. The losers, if the contractor does decide to take the clause literally, will be the employer who will foot the bill for compensation event claims under clause 60.1(7), and the project manager who will be run off his feet.

For practical reasons it may well be appropriate for the project manager to delegate his powers under clause 73.1 to the supervisor.

### Clause 73.2 – title to materials

The NEC addresses directly a point of some uncertainty in many contracts – who owns materials taken off the site? If the phrase 'take away and dispose' is found in the specification or method of measurement, it is reasonable to assume that the contractor is given title and is entitled to retain the proceeds of any sale. In other cases there can be uncertainty.

Clause 73.2 of the NEC settles the matter by providing that the contractor has title to materials from excavation and demolition only as stated in the works information. It follows from this that if the works information is silent on title to materials (and it is a point which can easily be missed) then title remains with the employer. That could cause complications in relation to the contractor's obligations to dispose of surplus materials; to the employer's liabilities in respect of such materials even after they have left the site; and to adjustments to the contract price in respect of any resale value.

In most cases giving the contractor title will be a better option than the employer retaining title. A general statement in the works information that the title of all surplus materials taken off site passes to the contractor will probably be sufficient to achieve this.

# Chapter 13

# Risks and insurances

## 13.1 Introduction

Section 8 of the NEC deals with the allocation of risks and the insurance of risks. These are far from straightforward matters. Risks and liabilities may seem easy enough to apportion on paper but the law reports are full of complex cases which expose the difficulties of drafting contractual provisions which readily identify the risk carrier when disaster strikes.

The case of *The National Trust* v. *Haden Young Ltd* (1993) which concerned the destruction of Uppark House by fire resulting from the negligence of subcontractors working on the roof, is a classic example. The court held, amongst other things, that under a JCT Minor Works form of contract a clause imposing an obligation on the employer to insure in the joint names of the employer and the contractor, neither expressly nor by implication included an obligation to insure in respect of subcontractors.

**Professional advice**

Given the difficulties it is understandable that project managers or other professionals with a technical background are sometimes inclined to leave insurance matters to others they regard as better equipped to look after the employer's interests. This is sound policy to the extent that a project manager or similar should always ensure that the employer receives the best professional advice. But in so far as a project manager may have, by the terms of his appointment, a general duty to advise the employer on all contractual matters, the project manager should not assume that the employer has of his own accord recognised and understood the obligations and implications of the contract. To do so is to invite a charge of negligence.

In the case of *William Tomkinson & Sons Ltd* v. *The Parochial Church Council of St. Michael in the Hamlet* (1990), again on a JCT Minor Works contract, it was held that an architect who failed to advise the employer of certain risks and the need to insure against them was in breach of his duty of care.

## 13.1 Introduction

**Risks generally**

Standard forms vary in the extent to which they identify and deal with particular risks. The employer's responsibility for his own property and its contents is one area of significant variance. But generally matters to be considered include:

- damage to the works prior to take over
- damage to the works after take over
- faulty materials and workmanship
- thefts and vandalism
- design defects
- damage to the employer's property
- damage to third party property
- consequential losses from damage
- injuries to the contractor's employees
- injuries to the employer's employees
- injuries to third parties.

Some of these can be grouped together for drafting purposes.

**Allocation of risks**

The policy which underlies most standard forms is that risks should be allocated to the party best able to control them. Thus take over of the works by the employer is usually seen as a watershed in respect of the works. Up to that time the contractor has care of the works and is generally responsible for damage, whereas afterwards the employer becomes responsible – subject to the proviso that the contractor is responsible for any damage he causes whilst remedying defects.

Responsibility for damage or injury to third parties usually follows the cause but damage to the employer's property is the employer's risk in some contracts.

The contractor is almost invariably responsible for the quality of work and carries the risks of faulty workmanship and materials. Responsibility for defective design generally falls on the party which undertook the design but that is not always the case.

**Excepted risks**

Excepted risks, or the employer's risks as they are called in some contracts, are those risks which are expressly excluded from the contractor's responsibility. Typically they include:

## 13.1 Introduction

- acts or omissions of the project manager, employer or his servants
- use or occupation of the works
- damage which is the inevitable or unavoidable consequence of the construction of the works
- war, riots and similar non-insurable events.

Broadly the excepted risks fall into three categories:

- fault or negligence of the employer
- matters under the control of the employer
- matters not the fault of either party.

The logic of the first two categories is obvious enough; the argument for the third category, where it applies, is that the employer is the party better able to carry the risk.

### Limitations on liability

Some contracts place limitations on the liability of the contractor to the employer for his acts and defaults. Such limitations, however, apply only between the contractor and the employer and they do not protect the contractor against third party claims.

### Insurances generally

Certain insurances are required by law, for example, motor insurances and employer's liability. In addition construction contracts invariably impose insurance requirements on one or both parties to ensure that funds are available to meet claims and to facilitate the completion of the works.

Some forms specify only the insurances which the contractor must carry. Other forms place obligations to insure on both parties.

### Common insurance provisions

The common insurance provisions of construction contracts are:

- the contractor is responsible for care of the works until completion
- the contractor must insure the works to their full replacement cost
- the contractor must indemnify the employer against claims for injury to persons or damage to property
- the contractor must insure against that liability.

## Other insurance provisions

According to the amount of detail in the insurance clauses of particular contracts, other provisions may cover:

- approval of insurers
- production of documentary evidence
- minimum levels of cover
- maximum levels of excess
- the employer's rights if the contractor fails to insure
- professional indemnity
- joint insurances.

## Professional indemnity for consultants

Employers who engage consultants as designers almost invariably require that they have professional indemnity insurance.

Where the contractor is responsible for design, either in-house or through consultants, it might appear on the face of it to be no concern of the employer's whether or not professional indemnity insurance is maintained. However, it is not always seen that way and it is not uncommon for such insurance to be required.

For cover where consultants are the designers there are two problems. Firstly, there is the point that professional indemnity insurance is usually on a claims made basis so that the cover is only effective in respect of the year in which the claim is made. Thus, once a policy lapses there is no cover for past work. Consequently a contractual requirement for such insurance needs to be drafted to ensure that cover is maintained for the legal limitation period rather than merely the construction period as for other insurances.

Secondly, there is the problem that the legal responsibility of a professional designer is limited at common law to the exercise of reasonable skill and care, and his professional indemnity cover is usually similarly limited. A claim on a fitness for purpose basis will have no access to such insurance.

## Contractor's in-house design

Contractors who undertake in-house design can insure against the negligence of their own designers. The cover is usually defined as being in respect of a negligent act, error or omission of the contractor in performance of his professional activities.

The need for such insurance arises because a contractor's all risks policy usually excludes design entirely or limits the indemnity to damage caused by negligent design to third party property or construction works other than those designed.

An ordinary professional indemnity policy does not cover the contractor against the problem of discovery of a design fault before completion. At that stage there is no claim against the contractor as there would be against an independent designer. To overcome this, contractors usually seek a policy extension giving first party cover. In effect this amounts to giving the construction department of the contractor's organisation a notional claim against the design department.

**Insurance terminology**

Insurance clauses in contracts often use phrases which are not particularly clear in themselves but which have particular meanings to insurers. For example:

- Subrogation
  This is the legal right of an insurer who has paid out on a policy to bring actions in the name of the insured against third parties responsible for the loss.

- Waiver of subrogation
  An agreement by one party's insurers to give up its rights against another party.

- Joint names
  Insurance in joint names provides both parties with rights of claim under the policy and it prevents the insurer exercising his rights of subrogation one against the other.

- Cross liability
  The effect of a cross liability provision in a policy is that either party can act individually in respect of a claim, notwithstanding the policy being in joint names. Without such a provision, liability between joint names is by definition not between third parties and is not covered.

- All risks
  An all risks policy does not actually cover all risks since invariably there will be exceptions. However, the effect of an all risks policy is to place on the insurer the burden of proving that the loss was caused by a risk specifically excluded from cover. In contrast, under a policy for a specified risk it is the insured who must prove that his loss was caused by the specified risk.

## 13.1 Introduction

**Risks and insurances under the NEC**

Whatever the contract, insurances should always be entrusted to specialists. With the NEC the novelty of its language makes specialist involvement even more imperative than usual.

Moreover, although the NEC does broadly allocate risks in accordance with conventional principles and requires insurance cover as normal, it has some differences from traditional construction contracts. In particular, the contractor's risks are defined by reference to those risks which are not detailed as employer's risks; the contractor's obligations for care of the works extend to the issue of the defects certificate and do not end at completion; the employer may acquire insurance obligations under the contract.

## 13.2 The employer's risks

Clause 80.1 of the NEC sets out the employer's risks. This is the key clause of Section 8 of the contract because the contractor's risks are not specifically detailed and are stated only in general terms, in clause 81.1, to be those not carried by the employer. Thus for the allocation of risk when an incident occurs, reference has to be made to the detail of clause 80.1.

**Clause 80.1 – employer's risks**

Clause 80.1 lists the employer's risks under six groupings which can be broadly described as:

- general – i.e. use of the works, unavoidable loss or damage, employer's fault
- loss or damage of employer supplied goods
- war, riots and similar non-insurable events
- loss or damage after take over
- loss or damage after termination
- additional risks as listed in the contract data.

These risks which would in most contracts be described as the excepted risks apply, as appropriate, to both care of the works and third party liabilities.

**General employer's risks**

Clause 80.1 commences by detailing a general set of employer's risks.

## 13.2 The employer's risks

These are claims, proceedings, compensation and costs payable due to various specific causes for which the employer accepts responsibility.

The first of these causes is:

> 'use or occupation of the Site by the works or for the purpose of the works which is the unavoidable result of the works'.

The wording here is anything but clear. The words used do not relate naturally to one another and the arrangement does not assist in producing an obvious meaning.

Perhaps it means simply this: that if claims arise from the use of the site for the construction of the works and the damage caused is unavoidable, then the employer meets such claims. Or put another way, the employer indemnifies the contractor in respect of use of the site.

It is unlikely that this particular employer's risk extends to the employer's use of the works; that is covered in a later item in the list of risks (loss or damage to the parts of the works taken over by the employer).

The second of the general causes is:

> 'negligence, breach of statutory duty or interference with any legal right by the Employer or by any person employed by or contracted to him except the Contractor'.

In short, the employer takes the risk of claims arising from his own negligence, breach or interference; or from that of others for whom he is responsible.

The third of the general causes is:

> 'a fault of the Employer or a fault in his design'.

### Loss or damage to employer supplied goods

This risk applies to plant and materials supplied by the employer or others on his behalf until the contractor has received and accepted them. It is a risk which in most circumstances is so obviously the employer's risk that it is not normally stated. But given the definition of plant and materials in the NEC, the risk as stated in clause 80.1 might extend well beyond normal circumstances and arguably to such things as materials on site for earthworks.

### Non-insurable events

As is customary, the employer takes the risk of loss or damage to the works from war and other generally non-insurable events. The list is conventional with only pressure waves obviously missing. Some contracts do include *force majeure* with a definition of *force majeure* which covers events beyond the control of the parties.

Note, in relation to the reference to 'strikes, riots and civil commotion not confined to the Contractor's employees', that by clause 26.1 the contract applies as if a subcontractor's employees are the contractor's.

**Loss or damage after take over**

This is an important employer's risk because clause 35.3 states that the employer takes over parts of the works when he begins to use them. Therefore this risk relates to the employer's use of the works both before and after they are completed.

It should be noted, however, that there are stated exceptions to the take over requirement in clause 35.3 and there are stated exceptions to the employer's liability in clause 80.1. One such exception which may attract some attention is that the employer may state a reason in the works information for not taking over parts of the works when he begins to use them. This could become a device whereby the employer does not take the risks of loss or damage to the parts of the works he has put into use.

The exception in clause 80.1 relating to the 'activities of the Contractor on the Site after take over' may also attract some attention. This is placed in the clause as though it indicates some fault on the part of the contractor. But note that by clause 82.1 the contractor is obliged to repair the damage to the works until the defects certificate, whether or not he is responsible. The contractor may for example be on site to repair vandal damage after take over. It is questionable whether in such circumstances the contractor should be responsible for loss or damage to the works unless he is in some way at fault.

**Loss or damage after termination**

This risk falls naturally on the employer once the contractor has left or been expelled from the site.

**Additional employer's risks**

It is not unusual, particularly in large projects or for projects where there will be many contractors on the site, for the employer to carry additional risks and to take out comprehensive project insurance. This can reduce the potential for disputes and it avoids duplication of insurance premiums. Additional risks are to be stated by the employer in the contract data.

## 13.3 The contractor's risks

The NEC avoids the usual practice of detailing the contractor's risks and stating the employer's risks as exceptions. It details the employer's risks and places all other risks on the contractor.

**Clause 81.1 – the contractor's risks**

Clause 81.1 is brief. It states simply that from the starting date until the defects certificate has been issued, the risks not carried by the employer are carried by the contractor.

The clause as drafted may cause some concern to contractors and their insurers in that the risks extend until the defects certificate, and any reduction of risk after take over or completion can only be determined from analysis of the employer's risks. This is significantly different from the approach of most standard forms.

## 13.4 Repairs

The manner in which contracts express the contractor's obligations to 'make good', 'maintain' or 'repair' the works after completion can be of the greatest importance if defects or damage arise.

In construction contracts the usual position is that the contractor has an obligation during the defects correction period or defects liability period (whatever it is called) to make good defects and damage for which he is responsible, and he has an entitlement to enter the site to do so. The burden of proof establishing the contractor's responsibility is normally with the employer.

In some process and plant contracts the contractor's obligation to make good defects and damage during the liability period is more widely drawn and the burden of proof may be on the contractor to show that he is not responsible.

The NEC appears to take the latter approach.

**Clause 82.1 – repairs**

Clause 82.1 provides that:

- until the defects certificate has been issued
- and unless instructed otherwise by the project manager
- the contractor promptly
- replaces loss of, and repairs damage,
- to the works, plant and materials.

Perhaps this clause would be better placed in Section 4 of the NEC with defects rather than in Section 8 with risks and insurance. But placed as it is, it carries two important provisions, the second of which follows from the first:

- the contractor has obligations to replace loss of and to repair damage to the works until the defects certificate is issued irrespective of the cause
- the risk lies with the contractor unless he can show that it is an employer's risk.

The clause is silent as to how and on what basis the contractor is to be paid for repair works which are not his responsibility. It may be that compensation event 60.1(14) is intended to apply, but procedures for assessing compensation events are hardly applicable after completion. There is in any event a potential timing problem in that a compensation event may not be notified after the defects date (clause 61.7), whereas the defects certificate can be issued after the defects date (clause 43.2).

## 13.5 *Indemnity*

Clauses 83.1 and 83.2 are conventional indemnity provisions.

**Clause 83.1 – indemnity**

Clause 83.1 provides simply that each party indemnifies the other against claims etc. due to an event which is his risk.

**Clause 83.2 – contributory reduction**

The liability of each party for his own risk is reduced in proportion to the other party's contribution to the event responsible for the claims etc.

## 13.6 *Insurance cover*

The NEC conveniently tabulates the insurance cover to be provided by the contractor. The core clauses put no express obligation on the employer to insure against his risks but the employer is required to provide insurance cover if he has stated in the contract data that he will do so.

## Clause 84.1 – provision of insurances

This requires:

- the contractor to provide insurances as stated in the insurance table, except insurances to be provided by the employer as stated in the contract data
- the contractor to provide additional insurances as required by the contract data.

## Clause 84.2 – the insurance table

The opening provisions of clause 84.2 are important:

- the contractor's insurances are to be in joint names
- the insurances are to cover events which are the contractor's risk
- the insurances are to give cover from the starting date to the issue of the defects certificate.

Note that because the contractor's insurances need only to cover his own risks, damage to the works from the employer's negligence is not covered by the insurances.

The insurance table provides that:

- insurance of the works, plant and materials is to be for replacement cost
- insurance of the contractor's equipment is to be for replacement cost
- third party cover is to be for the amount stated in the contract data for any one event with cross liability
- cover for the contractor's employees is to be for the greater of the amount required by the applicable law or the amount stated in the contract data.

## 13.7 *Insurance policies*

Clauses 85.1 to 85.4 deal with the details for inspection, approval and compliance with insurance policies.

## Clause 85.1 – submission of policies

This clause requires the contractor to submit his policies and certificates of insurances to the project manager for acceptance:

- before the starting date, *and*
- afterwards as the project manager instructs.

The only stated reason for not accepting the policies and certificates is that they do not comply with the contract. Rejection for any other reason will be a compensation event – clause 60.1(9).

### Clause 85.2 – waiver of subrogation

Insurance policies (and this would seem to include the employer's stated insurances as well as the contractor's) are required to include waiver of subrogation rights by the insurers against directors and employees of all insured.

### Clause 85.3 – compliance with policies

The parties are required to comply with the terms and conditions of the insurance policies.

### Clause 85.4 – unrecovered amounts

This clause makes the point that amounts recovered from insurances do not act as limits of liability. It states that amounts not recovered are borne by the parties causing the risk.

## 13.8 Contractor's failure to insure

### Clause 86.1 – contractor's failure to insure

Clause 86.1 provides that the employer may:

- insure a risk which the contractor is required to insure
- if the contractor does not submit the required policy or certificate.

The cost of the insurance to the employer is to be paid by the contractor.

## 13.9 Insurance by the employer

Clauses 87.1 to 87.3 provide for the contractor to have similar rights in respect of insurances to be provided by the employer, as the employer has in respect of insurances to be provided by the contractor.

### Clause 87.1 – submission of policies

Clause 87.1 matches clause 85.1 in requiring the project manager to submit the employer's policies and certificates to the contractor for acceptance.

### Clause 87.2 – acceptance of policies

Clause 87.2 qualifies the significance of acceptance by the contractor of the employer's policies and certificates by stating that acceptance does not change the responsibility of the employer to provide insurances as stated in the contract data.

### Clause 87.3 – failure by the employer to insure

Clause 87.3 matches clause 86.1 in permitting the contractor to insure in the event that the employer fails to insure, the cost in this instance being paid by the employer to the contractor.

# Chapter 14

# Disputes

## 14.1 Introduction

The most extensive changes between the first and the second editions of the NEC were made in the clauses of Section 9 covering the settlement of disputes. The adjudication provisions in those clauses were of strictly limited application and that had a correspondingly restricting effect on the matters which could be referred to arbitration. That in turn had the wholly unintended effect of apparently directing disputes towards litigation.

The second edition of the NEC remains firmly committed to the concept of adjudication and now requires all disputes to be referred to adjudication as the first stage of the dispute resolution process. The second stage process is not necessarily arbitration; it is described as 'review by the tribunal'. The nature of the tribunal is to be identified in the contract data. It could be arbitration, litigation or a disputes resolution panel.

The dispute provisions in the second edition are without any doubt an improvement on those in the first edition, but they remain nevertheless a subject of fierce debate. One eminent construction lawyer has publicly said of them that they do not lead to any readily enforceable results, are unlikely to lead to harmonious dealings between the parties, and are probably ineffective in the context of the English legal system.

The debate focuses on three major issues. The first is whether adjudication as intended in the NEC goes beyond the proper scope of adjudication; the second is whether the procedural requirements of the NEC are so restrictive as to be self-defeating; the third is whether the adjudicator's powers extend to opening up and reviewing decisions of the project manager.

## 14.2 The nature and scope of adjudication

Unlike arbitration, there is no statutory definition of adjudication (at the time of writing this book) and very little legal authority on its control or enforcement. This raises the question of whether a person acting as an adjudicator could, in law, find himself inadvertently acting as an arbitrator.

## 14.2 The nature and scope of adjudication

Section 32 of the Arbitration Act 1950 defines an arbitration agreement as 'a written agreement to submit present or future differences to arbitration, whether an arbitrator is named therein or not'. The courts appear to take the view that if a third party is appointed to settle disputes, the capacity in which he acts is a matter of construction and not one of description. See the cases of *Langham House Developments Ltd* v. *Brompton Securities Ltd* (1980) and *Cape Durasteel Ltd* v. *Rosser and Russell Building Services Ltd* (1995).

The courts do recognise a difference between an 'expert' and an 'arbitrator'. An expert determines variables, such as quantum, but his capacity does not extend to determining legal issues as does an arbitrator's. An expert who acts outside his limited capacity may well be acting as an arbitrator.

Providing that the capacity of an adjudicator is confined to that of an expert, there should be no confusion of his role with that of an arbitrator. But if, as is the case in the NEC, the adjudicator is required to settle 'any dispute arising under or in connection with the contract', then it is arguable that he settles legal issues as an arbitrator and not as an adjudicator.

The consequences of this, should it be the case, are that the 'adjudicator' would be subject to the control of the courts and liable to removal for misconduct for any departure from the rules of natural justice. And so fundamental is the point on whether adjudicators can settle points of law, that some eminent lawyers argue that any contractual provision which purports to confer the power to do so is invalid – since the parties have no power to agree to oust the jurisdiction of the courts. And any adjudication in such circumstances would be devoid of legal effect.

The point is made above that at the time of writing this book (January 1996) there is no statutory control of adjudication. However, this may change if the Housing Grants, Construction and Regeneration Bill is passed into law later in 1996, as is expected. That bill, as presently drafted, gives a party to a construction contract the right to refer a dispute arising 'under the contract' to adjudication.

## 14.3 The procedural requirements of the NEC

The main problem with the procedural requirements for adjudication in the NEC is that they can arrive at a stage of finality before the matters in dispute have been properly examined and considered. The effect could be that an assertion by one party could become final and binding (until revised by the tribunal) simply because the opposing party has had no opportunity to present its defence. It would then apparently be for the opposing party to raise the matter before the tribunal – with the possibility that the burden of proof might then transfer from the original claimant to the respondent.

This situation comes about because the adjudicator is only permitted to consider information provided within four weeks of the submission to adjudication, unless the parties otherwise agree. The party submitting the dispute to adjudication can itself submit information up to the end of the four week period. It can then decline to agree to any extension of the period for the other party to prepare its reply.

The proper course for the adjudicator in such circumstances would be to decline to give a decision. But since, in any event, the adjudicator only has four weeks from the end of the period for providing information to notify his decision, and that four weeks would rapidly be absorbed by exchanges between and with the parties, so far as that submission to adjudication is concerned, the adjudicator would become *functus officio* and would have no power to give a decision. The effect of that would be that the adjudicator would have failed to settle the dispute and the parties would be obliged to proceed as if the disputed matter was not disputed (clause 90.2) – subject only to the disputed matter being referred to the tribunal after completion.

The possibility of this sequence of events occurring is by no means remote, and it would not necessarily occur only by intent. The fact is that the four week period for submitting information is hopelessly inadequate for exchanges of case in a complex dispute, but there is no incentive for the claiming party to agree to an extension if its claim can be established by not doing so.

Intending users of the NEC, employers in particular, would be well advised to consult their lawyers on how to deal with this procedural difficulty.

## 14.4 The power to review decisions

Comment is made in Chapter 6, section 2 on the meaning of the word 'actions' as used in the NEC.

Under clause 90.1 the contractor can refer disputes on 'actions' or 'inactions' to adjudication and under clause 92.1 the adjudicator has power to revise and review any action or inaction. But the question is, what is the action in relation to a certificate or a valuation? Is it the process of certifying or valuing or is it the outcome of the process?

For the purposes of the NEC the answer is critically important. If an action is the process and not the outcome of the process then the project manager's decisions are not open to review by the adjudicator. And that suggests that they may be final and binding on the parties.

For clues in the contract as to what the answer might be, note that clause 93.2 permits the tribunal to review and revise 'any decision of the Adjudicator and any action or inaction of the Project Manager'. And note that only the contractor may submit disputes to adjudication on actions or inactions, so if actions or inactions include decisions the employer but not the contractor is bound by them.

## 14.5 Settlement of disputes

### Clause 90.1 – submission to adjudication

The opening sentence of clause 90.1 is intended as the agreement between the parties that all disputes shall be submitted in the first instance to adjudication. It provides that:

- any dispute
- arising under or in connection with the contract
- is submitted to and is settled by the adjudicator.

Clearly the sentence cannot be taken literally since the contract only governs disputes between the parties. It does not bind third parties with disputes connected with the contract.

The opening words of the sentence, 'Any dispute', appear to confer on the adjudicator unlimited capacity but there are difficulties in this as shown in section 14.2 above – particularly as the phrase 'arising under or in connection with' the contract suggests that it includes legal issues.

But it is not only the question of dispute on legal issues which presents a difficulty; there is also a difficult question on the timespan of the clause and whether it extends to disputes on latent defects which might arise many years after completion. The wording is general enough to suggest that it does, but the NEC adjudicator's contract in clause 6.3 states that the adjudicator's appointment is terminated when no further disputes 'under' the contract can be referred to him. It is not wholly clear when that is but it may be fixed by the defects date – see clause 61.7 on notifying compensation events. But, in any event, note the distinction between the words 'under or in connection with' used in clause 90.1 of the NEC and the word 'under' used in clause 6.3 of the adjudicator's contract.

On balance it seems unlikely that disputes on latent defects are intended to be submitted to adjudication – the quick settlement advantages of adjudication having long since expired – but the point is not certain.

A further point of interpretation on the opening sentence of clause 90.1 comes from the phrase 'is submitted'. This is almost certainly used to indicate that adjudication is a condition precedent to any dispute being referred to the tribunal. Thus, following the general style of the NEC it suggests that the parties have an obligation to submit any disputes between them to adjudication rather than a right to do so.

## Meaning of dispute

As to what is legally 'a dispute' and when it arises, see the comment by Lord Justice Steyn in the case of *M J Gleeson Group v. Wyatt of Snetterton Ltd* (1994): 'The ordinary meaning of 'dispute' in an arbitration agreement prima facie comprehends the case where a claim has been put forward and rejected'.

## The adjudication table

The adjudication table in clause 90.1 sets out which party may submit disputes to arbitration and the timescale for doing so. Note that in the table the phrase 'may submit' is used in contrast to the phrase 'is submitted' in the introductory sentence of the claim.

For disputes about actions or inactions of the project manager or the supervisor only the contractor may submit disputes to adjudication. The employer it would seem is bound by their actions or inactions. The timing requirement is that the contractor may only submit a dispute to adjudication between two and four weeks after notification of the dispute has been given to the project manager. The lead time of two weeks is presumably to allow a brief period for negotiations. There is however a further qualification. The notification (which presumably means the notification of the dispute) must itself be made within four weeks of the contractor becoming aware of the disputed action or inaction. Thus the latest time for submission to adjudication is eight weeks after the contractor becomes aware of the disputed action or inaction.

There may be some scope for manoeuvre on the timing by arguing what is meant by when the contractor 'becomes aware', but otherwise the pace at which the contractor is propelled towards making his submission to adjudication is not something which fits easily in a contract designed to reduce conflict. However, it is known that in some NEC contracts let to date common sense has prevailed and the parties have, by agreement, abandoned the timescale requirements for adjudication to give themselves a sensible time to consider the matters in dispute.

For disputes described as 'any other matter' (and which are not disputes about actions or inactions) either party may submit to adjudication, having notified the other party and the project manager of the dispute. Again the submission must be not earlier than two weeks nor later than four weeks after the notification of the dispute.

For those disputes on any other matters, the difficult question to assess is whether it is intended that the parties should have as much time as they like before deciding formally that they have a dispute which they wish to put to adjudication, or whether it is intended that as soon as one party rejects a claim or assertion made by the other, the clock starts ticking and

## 14.5 Settlement of disputes

after four weeks unless a submission to adjudication has been made, it is too late to mount a challenge to the claim or assertion. If the latter is the case then again the timescale is proactive.

**Clause 90.2 – the adjudicator's decision**

Clause 90.2 contains three provisions:

- the adjudicator is required to settle the dispute by notifying the parties and the project manager of his decision, together with reasons, within the time allowed by the contract
- unless and until there is a settlement the parties and the project manager proceed as if the disputed matter were not disputed
- the adjudicator's decision is final and binding unless revised by the tribunal.

All three provisions have great importance in determining the rights of the parties but all three leave many unanswered questions.

With regard to the first provision the questions are whether the adjudicator's decision must necessarily resolve the matter in dispute and what are the consequences of a decision being given out of time. In short, is an adjudicator bound like an arbitrator to decide on all the matters put before him, or can his decision be that he has not sufficient information to make a decision on all the issues. As to a decision given out of time, the probability is that it is of no contractual effect or at least it can be treated so if one party chooses to take that approach.

The second provision that unless and until there is 'such a settlement' the parties proceed as if the disputed matter was not disputed, appears to negate the effectiveness of any negotiated settlement but this can hardly be what is intended. But more importantly it has the effect of validating any claim which is not settled by the adjudicator within the required timescale. It also suggests that the key factor which determines how the parties and the project manager proceed until there is a settlement, depends upon how the dispute is formulated. Clearly it is better to be the asserting party than the disputing party.

A lesser but potentially interesting aspect of this provision is that although it binds the project manager and the parties, it does not bind the supervisor whose actions or inactions may be the subject of the adjudication.

The third provision of clause 90.2 makes any decision given by the adjudicator final and binding unless and until revised by the tribunal. Note that under clause 93.1 the dissatisfied party has only four weeks to refer the decision to the tribunal.

The question this provision inevitably raises is that of enforceability.

Will the courts enforce the adjudicator's decision in the face of opposition from one of the parties? Clause 92.1 attempts to answer this question by stating that the adjudicator's decision is enforceable as a matter of contractual obligation. But that may not be sufficient to stop a party who is sued on the basis of non-compliance with an adjudicator's decision from raising a variety of defences relating to the conduct or the content of the adjudication.

## 14.6 The adjudication

Clauses 91.1 and 91.2 of the NEC lay down rules as to the conduct of the adjudication. Surprisingly, and somewhat regrettably, clauses 2.1 to 2.5 of the NEC adjudicator's contract also lay down rules – and the two do not match.

**Clause 91.1 – the timescale of adjudication**

The provisions of clause 91.1 can be summarised as follows:

- the party submitting the dispute to adjudication includes with his submission information to be considered by the adjudicator
- any further information from either party is to be provided within four weeks of the submission to adjudication
- the adjudicator is to notify his decision within four weeks from the end of the period for providing information
- the four week periods may be extended if requested by the adjudicator and agreed by the parties.

The implications of the procedural requirements in this clause have been considered in section 14.3 above. Further points which need to be noted are that there is no provision for discovery of documents or a right to be heard. The procedure reads as though it operates on a documents-only basis but it does not expressly prohibit the adjudicator from instigating adversarial or inquisitorial proceedings.

**Clause 91.2 – joinder of subcontract disputes**

Clause 91.2 permits the contractor to submit a subcontract dispute to the main contract adjudicator with the main contract dispute if the two are related. That does, of course, raise questions as to the jurisdiction of the adjudicator over the subcontractor. And whilst that jurisdiction might be conceded in a back-to-back NEC form of subcontract it might well be disputed elsewhere.

*14.6 The adjudication* 255

But quite apart from jurisdictional problems there can be serious timescale difficulties in joinder proceedings, particularly as the main contractor may find himself fighting actions which are similar but not identical on two fronts simultaneously and documents from one dispute need to be carefully considered before they are put forward as information in the other dispute.

## 14.7 The adjudicator

Clauses 92.1 and 92.2 detail how the adjudicator is to act and how a replacement adjudicator is to be appointed if that proves necessary.

**Clause 92.1 – the adjudicator**

The content of clause 92.1 is extensive. It provides:

- the adjudicator settles disputes as an independent adjudicator and not as an arbitrator (for comment on this see section 14.2 above)
- the adjudicator's decision is enforceable as a matter of contractual obligation and not as an arbitral award (for comment on this see section 14.5 above)
- the adjudicator's powers include power to review and revise any action or inaction of the project manager or the supervisor (for comment on this see section 14.4 above)
- any communication between a party and the adjudicator is communicated to the other party (this should ensure that both parties are aware of the information provided to the adjudicator and on which he will make his decision; however, it does raise questions as to whether the adjudicator is permitted to meet the parties privately)
- the adjudicator makes any assessment of additional cost or delay in the same way that a compensation event is assessed.

**Clause 92.2 – appointment of a new adjudicator**

Clause 92.2 deals with the appointment of a new adjudicator when the current adjudicator resigns or is unable to act. It does not expressly deal with the situation which is contemplated in clause 6.1 of the NEC adjudicator's contract whereby the parties by agreement terminate the appointment of the adjudicator. Presumably in that case the appointment of the new adjudicator is also by agreement.

The provisions of clause 92.2 can be summarised as follows:

- if the adjudicator resigns or is unable to act, the parties jointly choose a new adjudicator
- if the parties cannot agree on a new adjudicator within four weeks the person stated in the contract data chooses the new adjudicator
- the new adjudicator is appointed under the NEC adjudicator's contract
- the new adjudicator has power to settle disputes submitted to his predecessor but not settled by him at the time of his withdrawal
- the date of his appointment is the date of submission to him of unsettled disputes.

The last of these submissions is oddly worded since the date of appointment of a new adjudicator will be a matter of fact and may have nothing to do with pre-existing disputes. But perhaps what it means is that where there are pre-existing disputes, time starts running for procedural requirements from the date of the new appointment.

## 14.8 Review by the tribunal

The NEC, as mentioned in section 14.1 above, does not contain any familiar arbitration clause but provides instead that the parties may refer their disputes to a 'tribunal' if they are dissatisfied with the decision of the adjudicator. The tribunal is an identified term of the NEC, not a defined term, so it is necessary for the nature of the tribunal to be named in the contract data.

Note that, in the contract data, if the named tribunal is arbitration there is provision for naming the arbitration procedure.

### Clause 93.1 – reference to the tribunal

The provisions of clause 93.1 can be summarised as follows:

- either party may refer a dispute to the tribunal if the adjudicator has notified his decision or failed to do so within the time allowed
- the notification of intention to refer to the tribunal must be given to the other party within four weeks of the adjudicator's decision or the date by which it should have been given
- the tribunal proceedings are not started before completion of the whole of the works or earlier termination.

The intention of the clause is almost certainly to make it a condition precedent to referral of a dispute to the tribunal that the adjudication process should have been exhausted. And by implication that reference of the dispute to the tribunal should be a condition precedent to, if not a substitute for, the parties' rights to litigate.

## 14.8 Review by the tribunal

However, the contract does not actually say either of these things and it is not certain how effective they would be if it did. The problem is knowing what status the courts would give to the adjudication provisions of the NEC if called upon to exercise their discretion to grant a stay in proceedings if an action was brought in the courts without regard to the dispute resolution procedures of the contract.

A comparable point came up in the case of *The Channel Tunnel Group Ltd v. Balfour Beatty Construction Ltd* (1993) where the House of Lords had to decide whether a stay should be granted when there was no immediate prospect of arbitration because the contract required disputes to be first referred to a disputes resolution panel. The reasons given in that case for the court exercising its discretion and granting the stay were:

- the parties were large commercial enterprises, negotiating at arms length with considerable experience of construction contracts and disputes arising under them
- the disputes resolution clause was carefully drafted and the parties had concluded that its provisions gave a clear practical advantage over having proceedings before the relevant courts in England or France
- parties making a choice of dispute resolution procedure must show good reasons for departing from it
- the grant of a stay was in accordance with the interests of the orderly regulation of international commerce.

Whether the NEC would be viewed as favourably by the courts remains an open question but it certainly is doubtful if the timescales for adjudication in the NEC would be well received by the courts.

As to an attempt by one party to commence tribunal proceedings (other than litigation) without first going to adjudication, that would be a jurisdictional matter for the tribunal to consider. The courts would only become involved if one party challenged the jurisdiction of the tribunal.

### Clause 93.2 – powers of the tribunal

The first sentence of clause 93.2 stating that the tribunal settles the dispute referred to it is intended, perhaps, as an indicator of the finality of the tribunal's decision.

The powers conferred on the tribunal include the power to review and revise:

- any decision of the adjudicator, *and*
- any action or inaction of the project manager or the supervisor.

The final provision of clause 93.2 stating that a party is not limited in the tribunal proceedings to the information, evidence or arguement put to the adjudicator, matches similar provisions in other contracts where there are conditions precedent to arbitration.

# Chapter 15

# Termination

## 15.1 Introduction

The circumstances by which any contract may come to a premature end can be broadly categorised as:

- termination – by agreement of the parties
- frustration – arising from events beyond the control of the parties
- determination – based on the default of one of the parties.

Most standard forms of contract have something to say on some or all of these matters. But no matter how simple or straightforward the wording appears to be, the reality is that ending a contract prematurely is rarely simple or straightforward. More often than not it is highly contentious, uncertain in its outcome and painful in its consequences. If it has to be done, ending a contract is a job for lawyers, not for laymen, not least because even the terminology used is a layman's nightmare.

**Terminology**

Some contracts use the phrase 'termination of the contract' to cover all types of premature ending; some use the phrase 'determination of the contract'; others refer to 'determination of the contractor's employment under the contract'. But the terminology itself is not decisive of the process as this extract from the eleventh edition of *Hudson's Building and Engineering Contracts* shows:

> 'A very varied terminology has been used both judicially and in commerce to describe the process by which a party, unilaterally and by his own action, brings a contract to an end before it has been fully performed either by himself or the other party. Thus forfeiture, determination, termination, renunciation, rescission (and even repudiation when applied to the action of the innocent party in ending the contract), have been variously used in the cases and elsewhere. In context the different descriptions should generally be regarded as synonymous, with no significant differences of consequential effect'.

The NEC uses only the phrase 'termination' and within that phrase it covers what is described above as 'frustration' and 'determination'.

**Termination at common law**

The ordinary remedy for breach of contract is damages but there are circumstances in which the breach not only gives a right to damages but also entitles the innocent party to consider himself discharged from further performance. This is usually a breach so serious that it goes to the heart of the contract. Sometimes it is called a 'fundamental' breach.

In such circumstances the innocent party has the legal right to terminate the contract at common law and is not reliant on a contractual right to terminate.

**Termination under contractual provisions**

To extend and clarify the circumstances under which termination can validly be made and to regulate the procedures to be adopted, most standard forms of contract include provisions for termination. Many of the grounds for termination in standard forms, however, are not effective for termination at common law. Thus failure by the contractor to proceed with due diligence, failure to remove defective work and subcontracting without consent are often to be found in contracts as grounds for termination. But at common law none of these will ordinarily be a breach of contract sufficiently serious to justify termination.

The commonest and the most widely used express provisions for termination relate to insolvency. Again at common law many of these are ineffective and even as express provisions they are often challenged as ineffective by legal successors of failed companies.

The very fact that grounds for termination under contractual provisions are wider than at common law leads to its own difficulties. A party is more likely to embark on a course of action when he sees his rights expressly stated than when he has to rely on common law rights. This itself can be an encouragement to error. Some of the best known legal cases on termination concern terminations made under express provisions but found on the facts to be lacking in validity.

In *Lubenham Fidelities* v. *South Pembrokeshire District Council* (1986) the contractor terminated for alleged non-payment whilst the employer concurrently terminated for failure to proceed regularly and diligently. On the facts, the contractor's termination was held to be invalid. But in *Hill & Sons Ltd* v. *London Borough of Camden* (1980), with a similar scenario, it was held on the facts that the contractor had validly terminated.

## Parallel rights of termination

Some contracts expressly state that their provisions, including those of termination, are without prejudice to any other rights the parties may possess. That is, the parties have parallel rights – those under the contract and those at common law – and they may elect to use either.

Other contracts, including the NEC, are silent on the issue but the general rule is that common law rights can only be excluded by express terms. Contractual provisions, even though comprehensively drafted, may not imply exclusion of common law rights.

The point came up in the case of *Architectural Installation Services Ltd* v. *James Gibbons Windows Ltd* (1989) where it was held that while a notice of termination did not validly meet the timing requirements of the contractual provisions, nevertheless there had been a valid termination at common law. However, in the recent case of *Lockland Builders Ltd* v. *John Kim Rickwood* (1995), the Court of Appeal held that an express term in the contract limited the scope of common law rights. It was said that although express termination clauses and common law rights can exist side by side, the common law right only arises in circumstances where the contractor shows a clear intention not to be bound by the contract.

## Termination under the NEC

The NEC puts its termination provisions in Section 9 of the core clauses together with the dispute provisions. Not too much should be read into this conjunction. It is not intended to imply that all disputes end in termination or that all terminations end in dispute. It is probably done simply to avoid the need for a Section 10 which would upset the clause numbering system of the contract.

Termination under the NEC can best be described as termination by numbers. The contract has 21 numbered reasons for termination, four numbered procedures for termination, and five numbered methods of calculating amounts due on termination. A table is included in clause 94.2 to show how the various reasons, procedures and amounts due relate.

But although the termination scheme of the NEC is elaborate and unusual in its presentation, it is not for the most part unusual in its content. However, what is unusual about the NEC scheme is that it permits the employer to terminate for any reason above and beyond the 21 numbered reasons. In other words, the employer has the contractual right to terminate at will; or, as some would say, the contract has a 'convenience' clause.

## 15.1 Introduction

**Legal effects of termination**

In construction contracts, termination is often expressed as 'termination of the contractor's employment under the contract', as though to emphasise that the contract itself is not terminated and that some of its provisions, particularly those for assessing amounts due and dispute resolution, remain in force.

In relation to arbitration provisions and provisions limiting liability for negligence, such wording is probably superfluous since those provisions survive independently of the main contract – see the House of Lords decision in the case of *Heyman* v. *Darwins* (1942). By contrast, provisions for liquidated damages may not survive termination – see the recent case of *Bovis Construction (Scotland) Ltd* v. *Whatlings Construction Ltd* (1994). For other provisions the wording of the contract is likely to be decisive.

The NEC in its termination provisions sets out in some detail certain procedures and assessments which survive termination. This, itself, probably reduces the scope for implying that other provisions should also survive.

## 15.2 *Summary of the NEC termination provisions*

Clauses 94 to 97 of the NEC deal with termination under the contract.

**Termination certificate**

Termination is commenced when the project manager issues a termination certificate – which he is obliged to do at the request of either party if the reason given complies with the contract (clause 94.1).

Once the termination certificate is issued the contractor does no further work (clause 94.5).

**Reasons for termination**

The stated reasons for termination include:

- insolvency of either party – reasons R1 to R10
- specified contractor's defaults – reasons R11 to R15
- non-payment by the employer – reason R16
- circumstances beyond the control of the parties – reasons R17 to R18
- prolonged suspension – reasons R19 to R21

In addition to the stated reasons the employer (but not the contractor) can terminate for 'any reason' (clause 94.2).

## Action on termination

On termination the employer is entitled to complete the works or employ others to do so. The contractor leaves the site either by instruction or by his own choice. The employer retains and may use any plant and materials to which he has title. The employer's rights to use the contractor's equipment depend upon the reason for termination and apply only where there is insolvency or other default by the contractor (clauses 94 and 96).

## Amount due on termination

Within 13 weeks of termination the project manager is required to assess and certify the amount due to or from the contractor (clause 94.4). The amount due is determined by reference to the reason for termination but in all cases it includes as a base amount, A1, which is, in simple terms, the valuation of the work at termination. To this is added or deducted according to the reason for termination one or more of the following:

- the cost of removing equipment
- the costs of completion
- either all or half of the fee percentage applied to the uncompleted work.

The minimum amount due applies to the contractor's insolvency or default and is the valuation at termination less the cost of completing. The maximum amount applies when the employer terminates for a reason not stated in the contract and is the valuation at termination plus the contractor's costs of removing his equipment plus the full fee percentage applied to the uncompleted work.

## 15.3 *Termination for 'any reason'*

The inclusion in the NEC of the provision in clause 94.2 permitting the employer to terminate for any reason is probably to ensure that the contract is acceptable to employers with a genuine need for a 'convenience' clause on state, security or exceptional commercial grounds. But the inclusion raises complex legal questions on which there is little legal authority.

If it is open to the employer to terminate for any reason he chooses then it appears that he is not bound, nor ever was bound, to see the works of the contract through to completion. So however capricious the reason or however unfair to the contractor, the employer is apparently not in breach of contract by abandoning the works or by ordering the contractor off the site and completing with another contractor.

If that really was the case the employer could terminate in order to get the works completed at a lower price by another contractor; could terminate for 'any reason' to avoid the confrontation involved in terminating on the grounds of contractor's default; or, and most objectionable of all, could terminate for 'any reason' to deprive the contractor of his opportunity of lawfully terminating the contract.

However, the probability is that the employer does not have the freedom which clause 94.2 suggests. Firstly, the requirement of clause 10.1 – that the parties shall act in a spirit of mutual trust and co-operation – may act as a constraint on the employer such that termination 'for any reason' is open to challenge as a breach of contract. Secondly, the law may impose a test of reasonableness on the employer's action. See, for example, the Australian case of *Renard Construction Ltd* v. *Minister of Public Works* (1992).

It may, perhaps, be argued that the NEC avoids unfairness to the contractor in that if the employer does terminate for 'any reason', the amount due to the contractor is equivalent to the amount he would receive as damages for breach of contract. However, such an argument overlooks the potentially adverse impact that termination may have on a contractor's reputation or his organisational arrangements and it is, in any event, far from sound on financial grounds.

The amount due on termination as calculated under the NEC may be considerably less than the contractor could claim as damages for breach of contract. The fee percentage, for example, is a figure obtained in competitive circumstances and it may not truly reflect the contractor's lost overheads and profit. Moreover, the contractor may have liabilities to subcontractors and suppliers which are not fully recoverable under the NEC's definition of 'actual cost'.

There is also the point that an innocent party may have two alternative claims in law for wrongful termination: one claim in contract as damages for breach, the other in *quantum meruit* for work done under the old rule established in the case of *Lodder* v. *Slowey* (1904). That rule, which operates only when the contractor is blameless, was expressed as follows by the New Zealand Court of Appeal prior to its affirmation by the Privy Council:

> 'The law is clear enough that an innocent party who accepts the defaulting party's repudiation of a Contract has the option of either suing for damages for breach of contract or suing on a Quantum Meruit for work done. An election pre-supposes a choice between remedies, which presumably may lead to different results. The nature of these different remedies renders it highly likely that the results will be different. If the former remedy is chosen the innocent party is entitled to damages amounting to the loss of profit which he would have made if the contract had been performed rather than repudiated; it has nothing

to do with reasonableness. If the latter remedy is chosen, he is entitled to a verdict representing the reasonable cost of the work he has done and the money he has expended; the profit he might have made does not enter into that exercise. There is nothing anomalous in the notion that two different remedies, proceeding on entirely different principles, might yield different results. Nor is there anything anomalous in the fact that either remedy may yield a higher monetary figure than the other. Nor is there anything anomalous in the prospect that a figure arrived at on a Quantum Meruit might exceed, or even far exceed, the profit which would have been made if the Contract had been fully performed.'

The advantage to the contractor of the application of the *Lodder* v. *Slowey* rule is that it enables him to escape from the prices in a poorly priced contract. It also acts as a restraint on the employer in preventing him from terminating when the contractor has performed underpriced early work but still has profitable later work in the contract.

Any employer contemplating terminating under the NEC for 'any reason' would be well advised, therefore, to take legal advice before proceeding on whether the amount due to the contractor as calculated under the contract really is the full extent of his financial liability.

## 15.4 Termination under clause 94 of the NEC

### Clause 94.1 – notification of termination

The first provision of clause 94.1 is that a party wishing to terminate should give notice to the project manager giving 'details' of his reasons for terminating. The second provision is that the project manager shall issue a termination certificate 'promptly' if the reason complies with the contract.

It is not clear if the reference to 'details' in the first provision means simply that the notifying party must specify which of the 21 numbered reasons is relied on or whether, in the case of the employer terminating for 'any reason', the employer must also give details of that other reason. A requirement to give details in the latter case would suggest that to do so has some contractual effect – but this does not appear to be the case. 'Any reason' is a specified ground for termination by the employer so it is not open to the project manager to refuse a termination certificate once an application is made. And 'any reason' other than the numbered reasons attracts exactly the same procedures and assessment whatever its details (clause 94.2).

The requirement in the second provision for the project manager to issue a termination certificate 'if the reason complies' puts a heavy burden on the project manager to act with absolute fairness and impartiality between the parties. The project manager cannot be seen as the agent of the employer in this matter. He may have to decide on merit between contesting notifications and he may well find himself in conflict with the employer and obliged to act against the interests of the employer.

In the event of the project manager refusing to issue a termination certificate because, in his view, the reason does not comply with the contract, the parties are probably expected to continue performance until the matter is resolved by adjudication.

If, nevertheless, one or both of the parties proceeds with the termination in the absence of a termination certificate, the probable effect is that all the termination procedures of the NEC are rendered ineffective and common law rules apply to any resulting dispute.

### Clause 94.2 – the termination table

Clause 94.2 limits the contractor's rights to terminate under the contract to the reasons listed in the termination table which forms part of the clause. The employer, however, is permitted to terminate under the contract for 'any reason'. For comment see section 15.3 above. The application of the termination table is straightforward.

### Clause 94.3 – implementation of termination procedures

Clause 94.3 states only that the procedures for termination are implemented immediately after the issue of the termination certificate. Note that clause 94.1 refers only to a certificate being issued 'promptly' and that there is no timescale in terms of days.

### Clause 94.4 – certification of amount due on termination

The project manager is required to certify within 13 weeks of termination (which is presumably the date on the termination certificate) the final amount due to or from the contractor.

There appears to be an assumption here that in the event of the employer's insolvency there will be sufficient funds available from one source or another to pay the project manager for his efforts. Or alternatively it may be thought that the project manager has acquired duties to both parties by his appointment and is obliged to fulfil those duties irrespective of the prospect of remuneration.

It is probably not worth speculating on the contractual position where there is no final certification by the project manager following the employer's insolvency, because in reality the contractor will lodge his claim with the receiver/liquidator regardless of whether or not he has a certificate.

One general aspect of clause 94.4 which is particularly worth noting is that certification of the final amount due after termination is not deferred until after the employer has completed the works and the costs of completion are known. Instead, the final amount due is based on an assessment made by the project manager – see amount A3 (clause 97.2).

### Clause 94.5 – cessation on termination

Clause 94.5 provides that after a termination certificate has been issued the contractor does no further work 'necessary to complete the works'. It may be arguable that work which is necessary for safety reasons is not work which is 'necessary to complete the works', but in any event the project manager, the employer and the contractor all have statutory obligations on safety and none can claim exemption by reference to contractual provisions.

## 15.5 Reasons for termination under the NEC

### Clause 95.1 – insolvency

Clause 95.1 details various financial failings in the nature of insolvency which are listed as reasons for termination, R1 to R10. Although separately numbered they attract identical treatment under the contract. Disappointingly the NEC includes in its list of financial failings, as indeed do most other standard forms, administration and arrangements with creditors, both of which are patently attempts to stay in business.

### Clause 95.2 – contractor's defaults

Clause 95.2 details three defaults which entitle the employer to terminate if the default is not rectified by the contractor within four weeks of notification:

- substantial failure to comply with obligations – reason R11
- not providing a required bond or guarantee – reason R12
- appointing a subcontractor for substantial work before the project manager's acceptance – reason R13.

It appears that the notification referred to in clause 95.2 is not the notice of termination referred to in clause 94.1 but is some earlier notice of dissatisfaction given to the contractor by the project manager. The intention is to put the contractor on notice of possible termination and to allow the contractor four weeks to rectify matters.

Reason R11 – substantial failure to comply with obligations – is exceedingly general and clearly the project manager's notice would have to be specific to be effective. As to what is meant by 'substantially failed' in reason R11, that is a matter of judgment on the facts having regard to legal precedents. The evidence required to support this reason should it be challenged would have to be convincing.

The second reason, R12 – failure to provide a bond or guarantee – is not uncommon.

The third reason, R13 – appointment of a subcontractor for substantial work before acceptance by the project manager – is hard to reconcile with the first part of clause 95.2 which allows the contractor four weeks to put the default right. Either the contractor has appointed a subcontractor before acceptance or he has not. Perhaps the clause means that the contractor has four weeks in order to obtain acceptance from being notified of his default.

**Clause 95.3 – contractor's continuing defaults**

Clause 95.3 details two defaults by the contractor of a continuing nature – or, at least, that is what the words 'not stopped defaulting' appear to suggest. The defaults are worded:

- substantially hindered the employer or others – reason R14
- substantially broken a health or safety regulation – reason R15.

It must be presumed that 'substantially hindered the employer or others' has something to do with the performance of the contract and is not a general complaint about the business activities of the contractor. As to how such hindering might occur, clause 25.1 (co-operation) stipulates the contractor's obligation to co-operate with 'others' and to share the working areas with them. Hindering 'others' could clearly affect completion of the project if not completion of the works. Hindering the employer is a more difficult concept to grasp. Hindering the employer in what? Premature use and take over of the works is one possible answer but is it conceivable that such a default attracts a harsher remedy than liquidated damages for late completion of the works?

Note that reasons R14 and R15 are grouped with reasons R11 to R13 in the termination table in clause 94.2. The distinction between the reasons in clause 95.2 and those in clause 95.3 appears to be therefore one of fact and not one of administrative significance.

## Clause 95.4 – failure to pay

Failure by the employer to pay within 13 weeks of the date of a certificate is one of the few reasons (R16), other than insolvency of the employer, entitling the contractor to terminate.

## Clause 95.5 – frustration

Clause 95.5 is not labelled as a frustration clause as such. But since there is no separate frustration clause in the NEC and the events detailed in clause 95.5 come close to what is normally described as frustration, it can probably be taken as the frustration clause of the contract.

At common law a contract is discharged and further performance excused if supervening events make the contract illegal or impossible or render its performance commercially sterile. Such discharge is known as frustration. A plea of frustration acts as a defence to a charge of breach of contract.

In order to be relied on, the events said to have caused frustration must be:

- unforeseen
- unprovided for in the contract
- outside the control of the parties
- beyond the fault of the party claiming frustration as a defence.

In clause 95.5 of the NEC the specified events which entitle either party to terminate are:

- war or radioactive contamination which has substantially affected the contractor's work for 26 weeks – reason R17
- events which under law release the parties from further performance – reason R18.

Note that the war reason (which is the only war clause in the contract) is not specifically related to war in the country of the works. So it is possible that far away war affecting materials required for the contract could be a valid reason for termination.

## Clause 95.6 – prolonged suspension

Prolonged suspensions of work are grounds for termination under most standard forms of contract. The NEC avoids the term suspension but clause 95.6 deals with what is normally termed prolonged suspension by detailing the circumstances which entitle the parties to terminate when the project manager's instruction to stop or not to start work has not been lifted within 13 weeks.

The rules are straightforward:

- the employer may terminate if the instruction is due to default by the contractor – reason R19
- the contractor may terminate if the instruction is due to default by the employer – reason R20
- either party may terminate if the instruction is due to any other reason – reason R21.

Reason R21 is what is known in some contracts as a *force majeure* clause. No fault is attributed to either party and the events which have caused the suspension are beyond the control of the parties.

## 15.6 Procedures on termination under the NEC

### Clause 96.1 – completion of the works

This clause provides firstly that the employer may complete the works himself or employ others to do so, and secondly that the employer may use any plant and materials to which he has title – procedure P1.

It is not wholly clear what purpose is served by a provision regulating the employer's conduct after termination. Any breach would seem to be outside the scope of the contract. As to plant and materials the employer is entitled to do whatever he wishes after termination with those to which he has title.

### Clause 96.2 – withdrawal from the site

Clause 96.2 deals principally with withdrawal from the site by the contractor and the subsequent use of the contractor's equipment. The clause states three procedures which apply according to the reasons for the termination:

- procedure P2 applies when the employer terminates for 'any reason' or the contractor is insolvent or in default. It entitles the employer to instruct the contractor to leave the site, remove equipment, plant and materials, and to assign the benefit of any subcontract or other contract related to the main contract to the employer

- procedure P3 applies when the contractor is insolvent or in default (or there is *force majeure*). It entitles the employer to use any equipment to which he has title
- procedure P4 applies when the employer is insolvent or in default (or there is *force majeure*). It entitles the contractor to leave the working areas and to remove his equipment.

The provision in clause 96.2 for assigning the benefits of subcontracts is similar to provisions found in many contracts. In practice, however, such provisions amount to very little since novation is normally required to form the contractual relationships which the clause envisages, and novations are concluded by agreement and not by compulsion.

For comment on the use by the employer of the contractor's equipment see Chapter 12, section 2.

## 15.7 Amounts due on termination under the NEC

### Clause 97.1 – valuation at termination

Clause 97.1 deals with the valuation of the work completed prior to termination and the costs incurred in expectation of completion. It fixes amount A1 due on termination by reference to five headings:

- an amount assessed as for normal payments. This will normally be the price for work done to date as clause 50.2 – but note clause 97.3 applying to Option A
- the actual cost for plant and materials brought within the working areas or for which the employer has title and has to accept
- other actual cost reasonably incurred in expectation of completing the whole of the works
- amounts retained by the employer
- unrepaid balances of advanced payments.

Amount A1 is the base for all amounts due whatever the reason for termination under the contract. It is much the same as the valuation at termination made under other standard forms of contract.

### Clause 97.2 – other amounts due

Clause 97.2 details the adjustments to the amount A1 (the valuation on termination) which are made to determine the final amount due on termination having regard to the reason for termination.

Four adjustment amounts are detailed:

- Amount A2 – the forecast actual cost of removing the contractor's equipment. This is included in the amount due to the contractor for all reasons except where the contractor is insolvent or in default.
- Amount A3 – the forecast additional cost to the employer of completing the whole of the works. This is deducted from the amount due to the contractor when the contractor is insolvent or in default.
- Amount A4 – the fee percentage applied to the difference between the tender total (or for Options E and F the first forecast of final actual cost) and the price for work done to date at termination. This is intended as a measure of the contractor's overheads and profit. It applies only when the employer has terminated for a reason not in the numbered list or when the employer is insolvent or in default.
- Amount A5 – this is half of amount A4. It represents a financial risk sharing between the employer and the contractor. It applies when neither party is responsible for the termination.

### Clause 97.3 – payment on termination – Option A

Clause 97.3 makes the necessary point that the amount due on termination is assessed without taking grouping of activities into account. To that extent it modifies how clauses 11.2(24) and 50.2 are to apply to clause 97.1.

### Clause 97.4 – payment on termination – Options C and D

Clause 97.4 applies to termination under the two target price contracts and it deals with how the contractor's share is to be calculated in the event of termination. It effectively fixes the calculation of the contractor's share to the price for work done to date at termination. This amount is then to be paid in accordance with clause 53 (the contractor's share).

# Chapter 16

# The NEC engineering and construction subcontract

## 16.1 Introduction

Many well used standard forms of main contract now have model forms of subcontract to complement a family of documents. The recent trend has been towards subcontracts which step down provisions from the main contract to the subcontract, so that as far as possible the obligations of the subcontractor are on a back-to-back basis with those of the contractor.

The NEC engineering and construction subcontract (the NEC subcontract) takes this trend to its limit in that it is precise duplication of the main NEC contract with little more than the names of the parties changed. It is, so far as contractual provisions go, a complete match of the main contract. It is, however, drafted in such detail that it is virtually independent of the main contract. It stands therefore as a subcontract which is fully back-to-back with the main contract but which, unusually, does not rely on examination of the terms of the main contract to give effect to its provisions.

That, at least, is the theory of the situation. In reality, however, with the NEC contracts much depends on the detail in the works information (subcontract works information for the subcontract) in fixing the obligations of the parties. So to obtain contracts which are back-to-back in both obligations and contractual provisions, it is as important to match the works information as to match the provisions.

**Use of the NEC subcontract**

Use of the NEC subcontract is not mandatory with the NEC main contract although its use is certainly encouraged by clause 26.3 of the main contract. If the contractor does not use the NEC subcontract he has to obtain the project manager's acceptance of any alternative conditions of subcontract.

To date there is little evidence from within the UK of any widespread use of the NEC subcontract, which suggests reservations amounting to reluctance on the part of main contractors to trade under its terms and conditions. The problem appears to be threefold:

- concern at the administrative burden
- a perceived imbalance in remedies for breach
- loss of traditional caveats and control mechanisms.

More is said on these points later in this chapter but if there is any useful generalisation to be made on why the NEC subcontract is apparently viewed with suspicion but other back-to-back subcontracts are in popular use, it may well be that the commercial instincts of main contractors inhibit experimentation with anything they cannot readily comprehend.

There is another possibility. If, as anecdotal evidence suggests, main contractors are learning quickly how to maximise their entitlements under the procedures of the NEC, they may not be particularly enthusiastic about affording the same opportunities to subcontractors.

**Structure of the NEC subcontract**

The NEC subcontract has the same structure as the NEC main contract with:

- core clauses
- main option clauses
- secondary option clauses
- schedules of cost components
- subcontract data – parts one and two.

With just a few exceptions the wording of the standard NEC subcontract documentation is identical to that in the main contract. And like the main contract, the subcontract places great reliance on non-standard documentation in works information, site information, contract data and programmes.

The basis of the transition of the standard documents from main contract documents to subcontract documents is that main contract references to the employer, the project manager and the supervisor, are replaced with references to the contractor; and references to the contractor are replaced with references to the subcontractor.

**Main option clauses**

The NEC subcontract repeats the five main options of the NEC, A to E, but has no equivalent of Option F, the management contract. Clearly it is not appropriate that the management function of Option F should be subcontracted.

It is for the main contractor to choose which of the main options should apply and his choice will not necessarily be governed by the main option applicable to the main contract. For example, the probability is that main contractors working under Options C or D (the target contracts) will seek to let as many subcontracts as they can under the firm price Options A and B, firstly to maximise their share of potential, and secondly to avoid the administrative burdens of the cost reimbursable arrangements of Options C and D. Even when the main contract is fully cost reimbursable under Option F, the probability is that the majority of the subcontracts will be let under Options A or B.

**Secondary option clauses**

The NEC subcontract has the same set of secondary options as the main contract and, with the exception of Option V, the trust fund, the wording matches one to the other.

Option V is significantly different because, instead of requiring the main contractor to establish a trust fund under the subcontract in like manner to the employer under the main contract, it simply draws attention to the employer's trust fund.

As with the main options it is for the main contractor to select which of the secondary options should apply to the subcontract. Again, there is no requirement that the two should match and the probability is that subcontracts will usually have a lesser number of secondary options than governing main contracts. The reasons for this will mainly be that for most subcontracts some of the secondary options will not be commercially appropriate or necessary; and for most subcontracts the full range of complexities in the main contract will not apply. See also the comment in section 16.4 below on damages for late completion.

**The core clauses**

Except for the changes in nomenclature described above and a few other changes in detail, the core clauses in the NEC subcontract are the same as those in the main contract. Accordingly the comment on the core clauses which follows in this chapter is not on the detail of the wording but on differences of application between subcontracts and main contracts.

## 16.2 Core clauses – general

**Actions**

The requirement to act in a spirit of mutual trust and co-operation applies in the subcontract by clause 10.1 and in subcontracts by clause 26.3.

## Communications

Much of the early general criticism of the NEC main contract has been that it generates too much paperwork and is costly to administer. If that is true of a main contract then the extension of the same communication requirements to a batch of subcontracts is something the main contractor will have to allow and prepare for as a serious matter.

## Early warning

Not all main contractors will readily take to the provision entitling the subcontractor to instruct them to attend early warning meetings.

## 16.3  Core clauses – the subcontractor's main responsibilities

### Design

Note that the employer's rights in respect of the subcontractor's design are preserved.

### Sub-subcontracting

The transfer of the full weight of the NEC main contract provisions to the subcontract seems heavy handed. Main contractors will not wish to find themselves burdened with considering, accepting or rejecting the terms of all sub-subcontracts, but that appears to be what is envisaged since it is most unlikely that many sub-subcontracts will be let under the NEC subcontract.

## 16.4  Core clauses – time

### Completion

The first question main contractors will have to decide in respect of time is whether or not to include secondary option R – liquidated damages for late completion. Most construction subcontracts leave damages for late completion unliquidated, not least because of the difficulties of making a genuine pre-estimate of loss when claims from other subcontractors may be a major element of any loss. Plant subcontracts, however, usually include liquidated damages as a means of limiting the liability of the subcontractor.

On a point of detail, note that the main contractor has two weeks to certify completion against one week allowed to the project manager in the main contract.

**Programmes**

The second question for main contractors, and a particularly difficult question, is what status to accord to programmes.

Main contractors are frequently torn between conflicting objectives. One is to tie the subcontractor down to a programme so that any departure by way of late completion of an activity is a breach of the subcontract entitling the main contractor to damages. The other objective is to allow themselves maximum flexibility to direct the timing of the subcontractor's activities, so that they are not in breach of the subcontract by preventing the subcontractor starting and finishing each activity as shown on the programme. For a recent case on the complexities of the situation see *Pigott Foundations Ltd v. Shepherd Construction Ltd* (1993).

Main contractors will almost certainly be concerned that the NEC subcontract gives them potentially the worst of both worlds in that the programme can form the basis of claims by the subcontractor under the compensation event rules, but there is no express corresponding liability on the part of the subcontractor for failure to perform to the programme.

**Take over**

Note that clause 35.3 is one of the four clauses of the NEC subcontract where the employer is mentioned. This clause deals with the employer's use of the subcontract works before completion and clearly it would not have been appropriate, in this clause, to replace the employer by the contractor.

## 16.5 Core clauses – testing and defects

For testing and defects the main contractor under the NEC subcontract assumes the roles of both the project manager and the supervisor, although in some clauses the jurisdiction of the supervisor is still expressly recognised.

**Uncorrected defects**

Comment was made in Chapter 9, section 8 on the implications of the liability for uncorrected defects of the main contractor under clause 45.1

of the main contract. The same liability transfers to subcontractors under the subcontract – namely for the assessed costs of having defects uncorrected rather than for the actual costs incurred.

With subcontracts this seems an even more repressive measure than with main contracts. Either the main contractor will correct the defect or he will not. In either case the financial implications can be properly ascertained. However, as clause 45.1 of the subcontract stands, the subcontractor can be liable to the contractor for costs which he may never incur or which exceed those actually incurred.

## 16.6 Core clauses – payment

Under the NEC subcontract the main contractor is required to assess and formally certify amounts due in like manner to the project manager under the main contract. However, there is some relaxation on the timing. Under the subcontract:

- the contractor has two weeks to certify (against one in the main contract)
- the contractor has four weeks from certification in which to pay (against three in the main contract).

The additional week allowed under the subcontract will obviously assist the contractor where the assessment dates in the subcontract and the main contract correspond. But by judicious timing of the subcontract starting date the contractor can gain more time.

### Retention

Contractors will need to be aware that retention only applies when secondary option P is included in the contract – and it is for the contractor to decide this irrespective of what is in the main contract.

## 16.7 Core clauses – compensation events

In subcontracts financial claims flow in both directions, whereas under main contracts financial claims from the employer are unusual, other than for liquidated damages or for the costs of remedying defects. This is in the nature of main contracts where the employer invariably has fewer stated entitlements to extra payments than the contractor.

Generally the intention of main contracts is that the employer should be able to liquidate his losses, defects apart. This is not the position in subcontracts and it is one of the objections to a straight transfer of main contract provisions into subcontracts as in the NEC subcontract.

The effect is that the subcontractor has numerous stated entitlements under the compensation event procedure to claim against the contractor for his breaches, but the contractor has little in return except liquidated damages or common law claims for damages.

**Timing requirements**

Presumably to ensure that the main contractor has time to pass on compensation event notifications where applicable, the subcontract has exceedingly tight timing requirements:

- the subcontractor has one week to give notice of a compensation event (against two in the main contract)
- the subcontractor has one week to submit a quotation (against three weeks in the main contract).

## 16.8 Core clauses – title

These clauses should operate much the same under the subcontract as under the main contract.

## 16.9 Core clauses – risks and insurance

**Risks**

The subcontractor's risks are defined as those which are not the employer's risks or the contractor's risks. The risks listed in the subcontract as the employer's risks and the contractor's risks match those listed as the employer's risks under the main contract. The arrangement looks clumsy and it may have the potential for confusion, not least because it directly involves the employer in risks under the subcontract.

**Insurance cover**

To avoid the expense of duplication of insurance cover it is not unusual for main contractors to arrange insurances which provide some cover to subcontractors. The NEC subcontract sets out a table of comprehensive insurance requirements but exceptions are allowed for any insurances which the contract data states are to be provided by the employer or the main contractor.

## 16.10 Core clauses – disputes and termination

### Adjudication

The subcontract adjudicator is not required to be the same person named as the main contract adjudicator, although there is nothing to prevent this and it could have advantages. It would certainly simplify the joinder provisions.

Note that the subcontract has two joinder clauses:

- clause 91.2 – which applies to the joinder of sub-subcontract disputes into a subcontract adjudication, *and*
- clause 91.3 – which applies to the joinder of subcontract disputes into a main contract adjudication.

The main contract has, of course, only one such clause.

Note also in clause 90.1 a small difference in detail between the two contracts. The subcontract requires notification of disputes about an action or inaction of the contractor within three weeks, whereas the main contract allows the contractor four weeks to give notice on actions or inactions of the project manager. This allows the main contractor a minimum of one week to pass on notice to the project manager.

One effect of the tight timescales for adjudication is that the subcontractor cannot be unduly delayed in having his dispute heard at least by an adjudicator. Subcontractors may see this as a significant improvement on the arrangements in some other standard subcontracts; main contractors may be concerned at being forced prematurely into a main contract adjudication in order to protect their position.

### Review by the tribunal

It is not essential that the subcontract adopts the same form of tribunal as the main contract, but arbitration is likely to be the usual choice. However, the subcontract has a clause (93.2), not found in the main contract, requiring that any subcontract dispute heard by the main contract adjudicator is referred to the main contract tribunal.

### Termination

Terminations are far more common under subcontracts than main contracts so the probability is that if the NEC subcontract comes into wide-

## 16.10 Core clauses – disputes and termination

spread use its termination provisions will soon be tested, particularly the provision in clause 94.1 permitting the main contractor to terminate at will.

One clause which will almost certainly attract some attention is clause 95.4 which states that the subcontractor may terminate if the contractor has not paid an amount he has certified within 13 weeks of the date of the certificate. This appears to bind the contractor to each certificate so that any later downward correction or set-off is an ineffective defence against termination.

Note, however, that failure to certify (as opposed to failure to pay on a certificate) attracts only interest under the subcontract (clause 51.4) and is not a stated ground for termination.

# Chapter 17

# The professional services contract

## 17.1 Introduction

The first edition of the NEC professional services contract (PSC) was published in September 1994 following the circulation of a consultative edition released in June 1992. Plans for a second edition are known to exist but consideration is being given by the promoters of the NEC to proposals from the Construction Industry Council for a harmonised set of conditions of engagement for employment of professionals in the construction industry and this will have some effect on when, or if, a second edition of the PSC is published.

**Intended usage**

The parties to the PSC are defined as the employer and the consultant. No restriction is put on the type of consultant and he may be the designer, project manager, supervisor, quantity surveyor or other provider of a professional service to the employer.

Nor is use of the PSC necessarily confined to professionals providing a service in respect of a NEC main contract. The PSC can theoretically be used with any form of main contract but there is little evidence as yet to suggest that this is happening.

**Style of the PSC**

The PSC follows closely the style of the NEC in language, layout and strategy and the policy is to have the same contractual framework for consultants as for contractors. Thus the PSC has nine sections of core clauses corresponding to those in the NEC and a batch of secondary option clauses, albeit fewer than in the NEC. It has four main pricing options and particulars of the contract are entered in contract data sheets parts one and two. It even applies to consultants similar provisions for programmes and compensation events as the NEC applies to contractors.

This similarity of treatment will not appeal to traditionalists in the professional fraternity and it may not appeal to all employers either. Some will question the wisdom of lining up their contractors and their consultants with common terms and conditions. But quite apart from questions of whether the similarities between the PSC and the NEC are potentially damaging to status and professional ethics, there is concern, already publicly expressed, that the PSC will generate administrative burdens on the consultant and the employer which will in the long term add to costs. And the point is made that whilst the employer employs a consultant (the project manager) to deal with administration under the NEC, he (the employer) may not be prepared or organised to undertake himself the burden of administration under the PSC.

**The brief**

The greatest concern, however, about the general style of the PSC is that it is no more than a set of general conditions and that it relies on the detail of the services to be provided being specified in an accompanying document – the brief. This document serves the same purpose in the PSC as the works information does in the NEC.

Numerous clauses of the PSC refer to information provided in the brief, so clearly unless the employer is a technically intelligent client he will have to employ a consultant to produce the brief. And if this consultant is employed under the PSC, on it goes ad infinitum.

**Guidance notes**

An official set of guidance notes accompanies the PSC and they will be invaluable to users of the contract. Note, however, that the guidance notes are not part of the PSC and they contain the warning that they should not be used for legal interpretation of its meaning.

## 17.2 The main options

The PSC has four main options each providing a different type of payment mechanism:

- Option A – priced contract with activity schedule
- Option B – time based contract
- Option C – target contract
- Option D – term contract.

It is for the employer to choose which main option to use although no doubt in many cases the consultants who are tendering or being considered for appointment will be invited to give their views on the most appropriate option for their particular services in the circumstances of the relevant project.

Surprisingly, perhaps, there is no main option for payment on a percentage fee basis.

### Option A – priced contract with activity schedule

This is a lump sum contract with the lump sum price broken down into individual lump sums for activities. It corresponds to Option A of the NEC and is suitable for use only where good definition of the detail and scope of the required services can be given at the outset.

As with any lump sum contract the financial risks on the adequacy of the contract price are with the consultant. The employer is assured of reasonable certainty of price providing that he has no changes of mind and commits no breaches of contract.

The function of the activity schedule in the PSC is much the same as that of the activity schedule in the NEC. In particular, it fixes entitlements to interim payments through the definition of the price for services provided to date (clause 11.2(15)). Although strictly the activity schedule is a document produced by the consultant and listed by him in part two of the contract data, the employer may well fix the activities by description leaving the consultant discretion only to distribute his lump sum price within the constraints of the schedule.

### Option B – time based contract

Option B of the PSC is a cost reimbursable contract corresponding to Option E of the NEC. The consultant is paid his actual cost on the basis of hours or days expended at tendered staff rates, plus a tendered percentage fee on the actual cost.

The only pricing risks to the consultant are in the adequacy of his rates and his fee percentage – which is to cover his overheads. The employer takes the risks on the amount of time spent in providing the services.

Expenses are paid either at rates itemised in the contract data or as part of the rates or the fee.

The reasons for using Option B of the PSC are much the same as those for using Option E of the NEC – see Chapter 2, section 7.

### Option C – target contract

Option C of the PSC corresponds to Option C of the NEC. The contract price is calculated by reference to the actual cost incurred and the application of a risk sharing formula to a tendered target price. For comment on this type of contract see Chapter 2, section 5.

Note, however, one significant difference between Option C of the PSC and Option C of the NEC. In the NEC interim payments to the contractor are based on actual costs 'paid' by the contractor; in the PSC interim payments to the consultant are based on actual cost of the services 'carried out' by the consultant (the reference in clause 11.2(16) of the PSC to the 'contractor' is presumably an error).

**Option D – term contract**

Option D of the PSC is described as a term contract. It applies to an agreement to provide services during a fixed period of time (the term). There is no equivalent main option in the NEC.

Term contracts can be quite complex arrangements in any walk of life for fixing the obligations of the parties. The key questions are: is the contractor entitled to and obliged to carry out all the work (or provide all the goods or services) required by the employer during the term? Or is the contractor simply on call (along with others) to meet the employer's needs as and when required, with each call either being an order under a single contract or an invitation to the contractor to enter into an individual contract?

The arrangement which appears to be envisaged in Option D of the PSC is that the consultant is obliged to provide such services as are ordered from a schedule of tasks during the term of the contract, but the employer is under no obligation to obtain all his services during the term from the consultant. The task orders are not intended as individual contracts.

## 17.3 The secondary options

The PSC lists eight secondary options for the employer to consider:

- Option E  – parent company guarantee
- Option F  – multiple currencies
- Option G  – transfer of copyright
- Option H  – employer's agent
- Option J  – termination at will
- Option K  – price adjustment for inflation
- Option L  – changes in the law
- Option M – special conditions.

This is a significantly smaller list than the list of secondary options in the NEC and accordingly the lettering of the options does not correspond.

However, Options E, F, K, L and M of the PSC are much the same in content as Options H, K, N, T and Z of the NEC. For comment see Chapter 3.

## Option G – transfer of copyright

Unless Option G is included in the contract, copyright of designs, documents etc. produced by the consultant remain with the consultant albeit that under clause 70 the employer is given certain rights of use. Option G, when included, provides for the transfer of copyright to the employer.

Clause G1 states that the copyright of 'documents' provided by the consultant and produced under the contract belongs to the employer. Clause G2 entitles the consultant to use 'designs' provided under the contract and to retain copies of 'documents' for his use. In both cases, the consultant is required to comply with any restrictions on use identified by the employer in part one of the contract data.

## Option H – employer's agent

The purpose of Option H is to allow the employer to appoint an agent to act on behalf of the employer under the contract. Clause H1 states simply that the employer's agent acts on behalf of the employer with the authority as set out in the contract data.

Use of this option is likely either where the employer is a large organisation and an individual is given delegated authority, or where the employer (perhaps a small organisation) relies on the appointment of an external consultant to manage affairs.

## Option J – termination at will

Clause J1 of the PSC is a particularly short clause even by the standards of the NEC. It states merely that the employer may terminate for any reason. The intention is clearly that when Option J is included in the contract the employer has the flexibility to terminate as he feels fit and irrespective of whether there is any default on the part of the consultant.

Arguably the use of this option is unnecessary since clause 93.3 of the PSC entitles the employer to terminate 'if the services are no longer required'. But depending upon how the word 'required' is interpreted there may be an implication in clause 93.3 that any transfer of services to another consultant would be a breach of contract.

Option J avoids the likelihood of a claim for breach of contract although it is questionable whether the financial arrangements at termination in the core clauses should apply when Option J is exercised.

## 17.4 Core clauses – general

The core clauses in section 1 of the PSC cover:

- clause 10 – actions
- clause 11 – identified and defined terms
- clause 12 – interpretation
- clause 13 – communications
- clause 14 – acceptance
- clause 15 – early warning
- clause 16 – ambiguities and inconsistencies
- clause 17 – health and safety
- clause 18 – illegal and impossible requirements.

These are much the same headings as in section 1 of the NEC but the numbering does not fully match because clause 15 of the NEC on adding to the working areas is omitted.

Note also that clause 14 of the PSC deals with acceptance, whereas clause 14 of the NEC deals with general obligations of the project manager and the supervisor.

### Clause 11.2(5) – the brief

The brief which is the key element of the PSC is defined in clause 11.2(5) as information which:

- defines the employer's objectives
- specifies and describes the services, *and is either*
- in the documents which the contract data states it is in, *or is*
- in an instruction given in accordance with the contract.

The requirement for the consultant to comply with instructions given by the employer changing the brief is in clause 21.3.

The clauses of the PSC which refer to the brief are as follows:

- clause 11.5 – definitions
- clause 13.4 – communications
- clause 21.2 – consultant's obligations
- clause 23.1 – co-operation
- clause 25.1 – approvals from others
- clause 30.3 – programme
- clause 32.1 – access to land and buildings
- clauses 40.1, 40.2 and 40.4 – quality management system
- clause 50.2 – assessing the amount due

- clause 70.1 — the parties' use of documents
- clause 72.1 — property provided by the employer
- clause 53.1 — Options B and C – actual cost
- Option F.1 — form of guarantee.

### Clause 11.2(6) – subconsultant

A subconsultant is defined as a person or corporate body who has a contract with the consultant to provide part of the services. The significance of the definition lies in clause 24.2 which requires the consultant to obtain the acceptance of the employer to the appointment of subconsultants.

### Clause 11.2(9) – defects

Clause 11.2(9) defines a defect as a part of the services which does not comply with the brief. As with the NEC, the definition of a defect has a narrow contractual meaning which implies fault on the part of the consultant in not complying with the brief.

### Clause 14.1 – acceptance

Clause 14.1 states that the employer's acceptance of a communication from the consultant does not change the consultant's liabilities or his responsibility to provide the services.

With the parties to the PSC in direct communication with one another, the reasonableness of this clause is questionable. Taken literally it could discourage or even negate constructive dialogue between the parties.

### Clause 15.1 – early warning

The early warning requirements of the PSC extend only to giving notice and, unlike the NEC, there is no express provision for early warning meetings. Failure by the consultant to give an early warning can affect the assessment of compensation events (clause 63.3).

### Clause 16.1 – ambiguities and inconsistencies

Clause 16.1 of the PSC suffers from a similar defect to clause 17.1 of the NEC in that the employer is required to give instructions resolving ambiguities and inconsistencies in any documents which are part of the contract – including, it would appear, those contributed by the consultant.

### Clause 18.1 – illegal and impossible requirements

Clause 18.1 has similar potential difficulties of application to clause 19.1 of the NEC. For comment see Chapter 6, section 11.

## 17.5 Core clauses – the parties' main responsibilities

### Clause 20.1 – the employer's obligations

The principal point of clause 20.1 is that it requires the employer to provide information in accordance with the accepted programme.

### Clauses 21.1 to 21.3 – the consultant's obligations

Clause 21.1 confirms only that the consultant's obligation is to provide services in accordance with the brief.

Clause 21.2 is more significant. It states that the consultant's obligation is to use reasonable skill and care in providing the services, except 'where stated otherwise'. There is an implication here that if the employer so specifies, the consultant may be obliged to provide services on a fitness for purpose basis.

Clause 21.3 empowers the employer to instruct changes to the brief and obliges the consultant to comply 'where it is practicable'. If it is not practicable the consultant is obliged to advise the employer of other measures which might be taken. It is not clear if the consultant is entitled to be paid for this.

### Clauses 22.1 to 22.2 – people

Where the consultant has named key people in the contract data, replacements are subject to acceptance by the employer.

### Clauses 24.1 to 24.3 – assignment and subcontracting

Clause 24.1 prohibits either party from assigning any benefit of the contract. Many employers will find this unacceptable.

Clause 24.2 requires the consultant to submit names of proposed subconsultants to the employer for acceptance. The definition of a sub-consultant in clause 11.2(6) is very wide and common sense is obviously required in the application of this clause.

## 17.6 Core clauses – time

The time provisions of the PSC are essentially programme related and they do not expressly cover obligations to complete the services in whole or in part within specified times. It is doubtful if failure by the consultant to comply with the accepted programme is a breach of contract entitling the employer to damages since there is no clear provision in the contract binding the consultant to the programme. This could, perhaps, be remedied, if thought necessary, by a statement in the brief that the consultant is to provide his services to the timing of an accepted programme identified in the contract data.

### Clauses 30.1 to 30.3 – the accepted programme

The accepted programme is either a programme identified in the contract data or a programme submitted within the period stated in the contract data. The programme is to include:

- the order and timing of the consultant's operations
- the order and timing of the work of the employer and others
- the dates the consultant plans to complete work needed to allow the employer and others to perform their work
- other information required by the brief.

### Clauses 31.1 to 31.3 – revising the programme

Clause 31.1 entitles the employer to instruct changes to the accepted programme.
 Clause 31.2 requires any revised programme to show:

- actual progress achieved
- the effects of compensation events and certified early warnings
- other proposed changes.

Clause 31.3 obliges the consultant to submit a revised programme for acceptance when instructed to do so and entitles the consultant to submit a revised programme when he chooses to.

## 17.7 Core clauses – quality

Section 4 core clauses deal with the consultant's quality management system and the correction of defects.

**Clauses 40.1 and 40.2 – quality management system**

Clause 40.1 requires the consultant to operate a quality management system which complies with the requirements for the services as stated in the brief. Clearly it is left to the employer to decide and specify in the brief the extent and the detail of the quality management system.

Clause 40.2 requires the consultant to submit for acceptance a quality policy statement and a quality plan.

**Clauses 41.1 and 41.2 – correcting defects**

Clauses 41.1 and 41.2 apply only to defects as defined in clause 11.2(9).

The consultant's obligation is to correct any defect which is non-compliant with the brief on becoming aware of it. In the event that the consultant does not correct such a defect, clause 41.2 entitles the employer to assess the cost of having the defect corrected by others. The consultant is then liable for the amount.

Because of the definition of a defect – non-compliance with the brief – this could be seen as a contractual remedy for the employer for almost any breach by the consultant. Applied in this way clause 41.2 could provide the employer with an alternative remedy to conventional damages for breach of contract.

## 17.8 Core clauses – payment

The PSC provides for interim payments by a similar mechanism to the NEC in that assessments are made at intervals as identified in the contract data. Note, however, that unlike the NEC there is no five week maximum interval.

**Clause 50.3 – the amount due**

Clause 50.3 fixes the amount due by reference to the price for services provided to date. This is a defined term which varies for each of the four payment options.

## 17.8 Core clauses – payment

**Clause 51.2 – payment**

The employer is required to pay within four weeks of receipt of the consultant's invoice or within such other period as stated in the contract data.

**Clauses 51.4 and 51.5 – interest on late payment**

Clauses 51.4 and 51.5 provide for compound interest on late payment. The rate of interest is as stated in the contract data.

## 17.9 Core clauses – compensation events

The compensation event procedure of the PSC follows the same general principles as the NEC. It identifies those matters, some of which are breaches of contract, for which the consultant can claim additional payment on an assessed actual cost basis.

**Clause 60.1 – compensation events**

Clause 60.1 lists seven compensation events:

- instructions changing the brief or the accepted programme
- failure by the employer to meet a date on the accepted programme for providing something
- failure by others to work within the times in the accepted programme or the conditions in the brief
- changes of decisions by the employer already given to the consultant
- requirements for the consultant to correct defects which are not his fault
- failure by the employer to reply to a communication from the consultant within the permitted period for reply
- withholding by the employer of an acceptance for a reason not stated in the contract.

**Clauses 61.1 to 61.5 – notifying compensation events**

The key points which emerge from clauses 61.1 to 61.5 are:

- either party 'may' give notice of a compensation event
- notice cannot be given more than two weeks after a party became aware of it

## 17.9 Core clauses – compensation events

- the employer may instruct the consultant to submit quotations for proposed changes
- the employer decides the validity of a compensation event claim
- the employer notifies the consultant of his decision within two weeks of notification
- if the employer accepts the claim he instructs the consultant to submit a quotation
- if the employer decides that there is too much uncertainty for an accurate quotation he states the assumptions on which the quotation is to be based.

The provision in clause 61.4 that if the employer decides that a compensation event is not valid it is 'deemed to have been withdrawn', repeats a provision found in the first edition of the NEC but omitted from the second edition. The intentions and the effects of the provisions are difficult to comprehend, not least because it is not for the employer to withdraw the consultant's claims.

### Clauses 62.1 and 62.2 – quotations for compensation events

The timescale for the submission by the consultant and consideration by the employer of quotations for compensation events is short, with only two weeks allowed for each.

### Clauses 63.1 to 63.4 – assessing compensation events

Clauses 63.1 to 63.4 fix the rules for assessments made by either the consultant or the employer.

Changes to prices for work already done are assessed using actual cost as defined in clause 11.2(11) and the fee. For work still to be done changes are based on forecast actual costs with allowance permitted for cost increasing and delaying factors which have a significant chance of occurring.

If the consultant has been notified that he did not give a requisite early warning, the savings which would have been made had proper warning been given are taken into account in assessing compensation events.

### Clauses 64.1 and 64.2 – the employer's assessments

The employer is expressly permitted to assess a compensation event if:

- the consultant has not submitted a quotation within the time allowed
- the employer decides that the consultant has not made a correct assessment.

This suggests that self assessment of additional costs by the consultant is intended as the normal rule.

It is not clear what the position is if the employer rejects the consultant's assessment but fails to notify his own assessment within the two weeks allowed. Arguably the consultant's assessment then becomes the valid assessment.

**Clauses 65.1 and 65.2 – implementing compensation events**

The important point of note here is the rule that assessment of a compensation event is not reviewed in the light of information which only becomes available after the consultant should have submitted his quotation. This is a similar rule to that in clause 65.2 of the NEC. For comment on the possible effects of the rule see Chapter 11, section 3.

## 17.10 Core clauses – title

Points to note in section 7 core clauses are:

- The employer may only use documents provided by the consultant after he has paid the consultant amounts due to date. Note that there is no suggestion of transfer of copyright to the employer and to achieve this secondary option G must be included in the contract.
- The parties are to treat information obtained in connection with the services as confidential. This presumably is intended only to prevent disclosure to third parties.
- Restrictions on publicising the services should be identified in the contract data.

## 17.11 Core clauses – risks and insurance

**Clauses 80.1 and 80.2 – the risks**

Clause 80.1 states the consultant's risks as claims etc.:

- for personal injury, death, loss or damage to property
- resulting from failure by the consultant to use reasonable skill and care.

Clause 80.2 states that the employer carries the risks for personal injury and death, and loss or damage to property which is not the consultant's.

These provisions should be treated with the greatest of caution as expressing the true legal position of the parties. It is far from clear if they

## 17.11 Core clauses – risks and insurance

are intended to apply only to third party liabilities or to liabilities generally.

### Clauses 81.1 and 81.2 – consultant's insurances

Clause 81.1 requires the consultant to provide insurance cover for:

- failure to use reasonable skill and care
- personal injury, death, loss or damage to property
- loss or damage to property provided by the employer for use by the consultant
- such other events as specified in the contract data.

In short the consultant must at the very least carry professional indemnity insurance, third party cover, and property insurance (if using the employer's property). He will, of course, also have to carry other statutory insurances in respect of personnel and vehicles.

### Clause 82.1 – limits of liability

Clause 82.1 limits the consultant's liability 'to the employer' to the amounts of insurance cover required by the contract or such other amounts as are stated in the contract data.

### Clause 83.1 – employer's insurances

The employer's contractual obligation to insure is limited to such insurances as are identified in the contract data. That, of course, says nothing of the employer's legal obligations.

## 17.12 Core clauses – disputes and termination

### Clauses 90.1 to 90.3

The PSC follows the NEC policy of requiring disputes between the parties to be referred to adjudication. The intention appears to be that adjudication is mandatory and is to be treated as a condition precedent to formal reference to arbitration (or some other dispute resolution tribunal). As with the NEC the timescale for adjudication is exceedingly short and likely to be inadequate for any complex dispute.

## Clauses 91.1 to 91.3 – the adjudicator

The adjudicator is intended to be appointed at the commencement of the contract and to remain in post to settle disputes until his appointment is terminated. For practical reasons this should probably be on completion of the provision of the services. The PSC is not clear on the point and arguably disputes on latent defects after completion have to be referred to adjudication.

Note that if it is necessary to appoint a replacement adjudicator, that adjudicator is to be appointed under the 'Institution of Civil Engineer's Adjudicator's Contract' – by which is probably meant the NEC Adjudicator's Contract.

The provisions for joinder of subconsultant disputes in clause 91.3 are, like most joinder provisions, well intentioned but likely to be highly problematic in their application.

## Clauses 92.1 and 92.2 – dispute resolution

The PSC, like the NEC, refers to a 'tribunal' rather than to arbitration as the final contractual resort in dispute resolution. However, if arbitration is the specified tribunal, the arbitration procedure can also be specified in the contract data.

## Clauses 93.1 to 93.3 – termination

Clause 93.1 permits either party to terminate in the event of the other becoming insolvent.

Clause 93.2 permits the consultant to terminate if the employer does not pay an overdue amount within four weeks of notice from the consultant that the payment is overdue. Since there is no certification process it is not clear how this provision works if there is a disputed invoice. Adjudication, which is the sensible remedy, should ideally be before termination and not after it.

Clause 93.3 permits the employer to terminate if the services are no longer required or if the consultant has substantially failed to comply with his obligations. Note that there are no stated procedures or notice provisions attached to the employer's termination. See the comment in section 3 above on the possible meaning of the phrase 'if the services are no longer required'.

## 17.12 Core clauses – disputes and termination

### Clauses 94.1 to 94.3 – procedures and payment after termination

After termination the employer is entitled to:

- complete the services himself or employ others to do so
- use any 'material' to which he has title
- require the consultant to assign the benefits of subconsultant contracts and the like.

The final payment due to or from the consultant includes:

- an amount as for normal payments
- the actual cost incurred in expectation of completing the services plus the fee
- any amounts retained by the employer.

For termination by the employer because of the consultant's insolvency or substantial failure to comply with his obligations, the final payment can include a deduction of the forecast costs to the employer resulting from the termination. This, of course, is normal practice but what seems to be missing is any provision for the consultant to recover lost overheads and profit when the employer terminates at his discretion without any fault on the part of the consultant; or the consultant terminates due to the employer's failure to pay.

# Chapter 18

# The adjudicator's contract

## 18.1 Introduction

The NEC includes in its family of forms an adjudicator's contract designed to take effect between the adjudicator and the parties to the dispute which has been referred to him. Those parties will be the employer and the contractor for a main contract dispute, the contractor and the subcontractor for a subcontract dispute, and the employer and the consultant for disputes under the professional services contract. Guidance Notes are issued with the adjudicator's contract.

The first edition of the adjudicator's contract was published in 1994 following the circulation of a consultative version in 1992. At the time of writing (April 1996) the first edition still remains in force notwithstanding the publication of second editions of other NEC documents in 1995. A second edition of the adjudicator's contract will be issued in due course but the timescale is dependent upon the success of efforts being made by various professional bodies and the promoters of the NEC to jointly produce a unified all-purpose adjudicator's contract.

In the meantime there are potential problems with conflicts of wording between the first edition of the NEC adjudicator's contract and the second editions of other NEC documents. The two are not compatible on the timing requirements for the submission of information. For further comment see section 18.5 below.

**Need for the adjudicator's contract**

Practitioners in the dispute resolution field, arbitrators, conciliators, mediators and adjudicators frequently operate with no more formality in the contract between themselves and the parties than agreement to their terms of appointment. A good reason for this is that it allows the dispute resolution process to be conducted with the maximum flexibility allowed by the law and any particular procedural rules imposed by the substantive contract.

The danger of formalising conditions of appointment in addition to terms is the uncertainty that this can introduce on the validity of awards,

## 18.1 Introduction

recommendations or decisions, or even on the validity of the proceedings themselves, in the event that allegations are made by one of the parties of breach of conditions. This is perhaps less important in adjudication than arbitration because adjudication lacks the legal standing of arbitration, particularly with regard to the enforceability of the outcome. Nevertheless it is still a potential inconvenience to the parties that adjudication proceedings under formal conditions may be vulnerable to breaches of what would normally be seen as relatively inconsequential matters.

The position of the adjudicator, if it is he who has breached the conditions of the adjudicator's contract, is also vulnerable. He may lose his entitlement to be paid for his services or even worse find himself liable to the parties for damages.

### Purpose of the adjudicator's contract

The general purpose behind the adjudicator's contract is to generate consistency and confidence in the operation of the adjudication provisions of the NEC. The procedural rules in the substantive contracts will go some way towards this but the huge numbers of adjudicators who will be needed if the NEC achieves its promotional ambitions could pose a problem if each adjudication appointment has its own individual terms and conditions. Adjudicators unlike arbitrators are appointed when contracts are made, not when disputes arise, so their numbers will be proportionally greater and their variety correspondingly more variable.

The purpose of the adjudicator's contract in specific terms is to set out the obligations and liabilities of the parties and the adjudicator on:

- actions
- communications
- submissions of information
- giving of decisions
- payments
- title
- risks
- termination.

### Use of the adjudicator's contract

Use of the NEC adjudicator's contract is not mandatory for NEC contracts except that the substantive NEC contracts, the Engineering and Construction Contract, the Subcontract and the Professional Services Contract, all provide that any replacement adjudicator shall be appointed under the NEC adjudicator's contract.

### Subcontract adjudication

The adjudicator appointed under the NEC Engineering and Construction Subcontract is not required to be the same adjudicator as appointed under the main contract. And it is probably best that he should not be, although it is not prohibited.

However, in respect of subcontract disputes which concern a matter disputed under the main contract, the contractor is entitled to have the subcontract dispute joined with the main contract dispute and referred to the main contract adjudicator.

## 18.2 Appointment of the adjudicator

The procedure for the appointment of an NEC adjudicator is not quite as straightforward as the NEC implies. Unlike the position in some other contracts which use adjudicators, it is not open to the employer under the NEC to decide and fix in the contract the name of the adjudicator. Unfortunately the entry space in part one of the contract data gives the impression that the employer names the adjudicator with the information he provides to tenderers.

Operated properly, all that the employer is entitled to do at tender stage is to put forward the name of a proposed adjudicator. The appointment itself must be subject to acceptance by both parties of the individual concerned and of his fees. And, of course, there must be no conflict of interest between the appointed adjudicator and the parties.

### Procedure for appointment

In practice employers usually approach prospective adjudicators before or during the tender period to ascertain their availability and fee scales. The Institution of Civil Engineers maintains a list of trained adjudicators, which is available on request.

Then, with the tender information or when the selection of the contractor is nearing completion, details of one or more proposed adjudicators will be released with the aim of obtaining conditional approval to the appointment of a particular individual. This is taken a stage further when the confidentiality of the tender process expires and potential conflicts of interest can be examined.

Finalisation of the appointment of the adjudicator normally only takes place after the substantive contract has been signed, not least because the adjudicator's contract is with the parties to that contract.

### Failure to agree on the adjudicator

The NEC does not contain provisions similar to those found in most arbitration clauses for an appointing body to be involved if the parties fail to agree on the appointment of an adjudicator. The full implications of this are difficult to assess but the point certainly highlights the desirability of the parties reaching agreement, at least in principle, on who the adjudicator should be before the substantive contract is awarded.

It is doubtful if the provisions in clause 92.2 for a person named in the contract data to choose a replacement arbitrator can be applied, without the consent of the parties, to a situation where an adjudicator has not been appointed in the first place. But the parties may have little option but to consent since to operate the NEC without an adjudicator would probably nullify the whole of the dispute resolution provisions of the contract.

### Replacement of an adjudicator

Clause 92.2 of the NEC has been considered in detail in Chapter 14, section 7. In short it provides that a person named by the employer in the contract data shall choose a replacement adjudicator, if the original adjudicator resigns or is unable to act and the parties fail to agree on a replacement. The parties are obliged to accept the choice so perhaps this puts on the person making the choice the burden of ensuring that his choice does not intend to charge excessive fees.

A replacement adjudicator is expressly required by clause 92.2 to be appointed under the NEC adjudicator's contract.

## 18.3 Joinder provisions

Clause 91.3 of the NEC Engineering and Construction Subcontract and clause 91.3 the Professional Services Contract both provide for joinder with main contract adjudications if there are related matters in dispute.

The arrangement has much to recommend it in theory since it aims to avoid conflicting decisions being given by different adjudicators on the same dispute. In practice, however, joinder can run into serious difficulties – not least because a contractor's disputes with subcontractors frequently have a domestic element which is of no concern to the employer.

As always with joinder there is the potential for unexpected conflict of interest, but this can be avoided under the NEC if the subcontractor can see from the subcontract data, as he should be able to, the name of the main contract adjudicator before the subcontract is made.

What is not clear under the NEC joinder arrangements is whether it is intended that the main contract adjudicator should have any formal recognition or written agreement in respect of his involvement in subcontract disputes. In the absence of such an agreement it is difficult to see how the adjudicator derives his authority in respect of the subcontract, or how the subcontractor can have any liability for a share of the adjudicator's fees. It is certainly arguable that a main contract adjudicator has no obligation to make a decision in respect of a subcontract unless he consents to do so. This is an issue which the parties and the adjudicator may wish to address on an ad hoc basis when the joinder provisions are invoked.

## 18.4 Section 1 – general

**Clause 1.1 – actions**

Clause 1.1 states that the parties and the adjudicator shall act as stated in the contract. The inference of this is that failure to do so is breach of the adjudicator's contract. The consequences of breach and the remedies for breach are stated only in respect of payments. The adjudicator himself appears to be immune since by clause 5.1 the parties indemnify him against claims arising out of his work.

No doubt when the adjudicator's contract is revised in some later edition the provision in clause 10.1 of the main contract, that the adjudicator shall act in a spirit of independence, will find its way into the adjudicator's contract. One would, of course, expect much more than a mere spirit of independence from an adjudicator and it is to be hoped that adjudicators do not take this to be the test of their proper performance.

**Clause 1.3 – definition of adjudication**

Clause 1.3 defines adjudication as:

- work carried out by the adjudicator
- on a dispute arising from the contract between the parties
- referred to him by either of the parties.

On a narrow interpretation the reference to the contract between the parties might seem to exclude the adjudicator's authority to involve himself in subcontract disputes. But if, on wider interpretation, a subcontract dispute is taken to be a dispute arising from the contract, the adjudicator's authority may extend to such subcontract disputes as are referred to him.

## Clause 1.4 – expenses

Expenses are defined as expenses incurred by the adjudicator, including payments to others consulted by him.

Clause 1.4, together with clause 2.4, appears to give the adjudicator a free hand to spend the parties' money on professional advisors however expensive. Some check obviously needs to be applied. Although it may be implied that the expenses should be both reasonable and reasonably incurred it would do no harm to say so and perhaps to add 'approved'.

## Clause 1.6 – interpretation

Clause 1.6 states that if 'another party' becomes a party to the dispute references to 'Parties' are interpreted as including the other party. This is presumably intended to give effect only to joinder provisions in the principal contracts but 'another party' is not defined and it may not necessarily be a party under the NEC regime.

## Clause 1.9 – retention of documents

By this clause the adjudicator is required to retain copies of documents provided to him by the parties for the period of retention stated in the contract data. There is no obvious mechanism for payment of any insurance costs or storage costs. And since clause 6.3 terminates the adjudicator's appointment after the reference of disputes, it is not clear how he can be burdened with a continuing obligation.

## 18.5 Section 2 – adjudication

### Clause 2.1 – submission of information

Clause 2.1 repeats the procedure in clause 90.2 of the first edition of the NEC. The corresponding clause in the second edition, clause 91.1, sets out a different procedure. The conflict can be eliminated by amending clause 2.1 or deleting it from the adjudicator's contract, so relying on clause 91.1 of the NEC for the procedural requirements.

### Clause 2.2 – further information

The entitlement in clause 2.2 for the adjudicator to ask for and to receive further information up to six weeks from the submission is not compatible with the provision in clause 91.1 of the NEC that information to be considered by the adjudicator must be provided within four weeks of the submission.

Again, it is probably easier to amend clause 2.2 or to delete it altogether than to adjust clause 91.1.

### Clause 2.3 – the adjudicator's decision

The requirement for the adjudicator to notify the parties of his decision within four weeks of receiving the information from the parties, or such longer period as may be agreed, does match the requirement in clause 91.1 of the NEC.

### Clause 2.4 – consultation

Clause 2.4 permits the adjudicator to consult others to help him make his decision. This is an entitlement which adjudicators should exercise with the greatest caution and arguably only after receiving consent from the parties. Adjudicators do not run the same risks as arbitrators of being removed for misconduct, but before an adjudicator consults others he should consider questions of confidentiality, conflicts of interest and cost.

## 18.6 Section 3 – payment

For adjudicators the great benefit of the adjudicator's contract is that in clauses 3.1 to 3.7 it details clearly the obligations of the parties on payment of his fees and expenses. The parties pay the fees and expenses in equal shares and payments are due within four weeks of the date of invoice after which they attract interest.

In the event of one party failing to pay, the other party becomes fully liable and is left with the task of recovering the amount due from the defaulting party. In short the parties are jointly and severally liable to the adjudicator for his fees and expenses.

The adjudicator's contract does not envisage the adjudicator following the practice of arbitrators and releasing his decision only after payment of his fees. To do so would probably be a breach of clause 2.3.

## 18.7 Section 4 – title

### Clause 4.1 – use of the adjudicator's decision

Clause 4.1 has two provisions:

- the parties may use the adjudicator's decision to resolve a dispute in connection with the contract
- the parties are to treat the decision and the information provided for the adjudication as confidential.

The first provision appears to be a statement of the obvious but it has deeper intent. Presumably it means that an adjudicator's decision can be used or relied on by the parties to resolve a dispute not referred to adjudication and that title to the adjudicator's decision rests with the parties and not the adjudicator.

The second question raises questions as to admissibility in legal proceedings involving other parties.

## 18.8 Section 5 – risks

Clause 5.1 has been much criticised by lawyers as being wholly inadequate for the purpose it is supposed to achieve.

### Clause 5.1 – indemnity

The clause states that the parties indemnify the adjudicator against claims, compensation and costs arising out of his work in connection with an adjudication.

The intention of this clause is supposedly to give an adjudicator similar protection from suit to that enjoyed by the arbitrator. Unfortunately, however, the wording of the clause is such that it does not protect the adjudicator against a claim for negligence in the performance of his duties. That could only be achieved by an express exclusion of liability for negligence and even then it would be subject to the reasonableness test of the Unfair Contract Terms Act 1977.

## 18.9 Section 6 – termination

### Clause 6.1 – termination by the parties

Clause 6.1 permits the parties to terminate the appointment of the adjudicator by notifying him of their agreement to terminate. The parties are not required to give reasons and it is unlikely that the adjudicator is entitled to any financial recompense other than payment of his outstanding fees and expenses.

The adjudicator is probably entitled, if not obliged, to return to the parties any documents provided by the parties which he has in his possession.

### Clause 6.2 – termination by the adjudicator

Clause 6.2 permits the adjudicator to terminate his appointment if he is prevented from carrying out his work as an adjudicator. The clause is slightly ambiguous in that it is not clear whether it means prevented by the parties (in the sense of obstructed) or prevented by personal circumstances. If it is the case that the adjudicator can resign if he finds himself busy or on extended holiday, the clause ought to have some minimum notice requirements to enable the parties to arrange the appointment of a replacement adjudicator.

### Clause 6.3 – termination on completion

Clause 6.3 provides that the adjudicator's appointment terminates when he has given his decision on all disputes referred to him and no further disputes can be referred to him.

For disputes on compensation events that is shortly after the defects date but for latent defects, the timescale can be much longer.

See also the comment on clause 90.1 of the NEC in Chapter 14, section 5.

# Table of cases

**Note:** End references are to chapter sections. The following abbreviations of law reports are used:

| | |
|---|---|
| AC | Law Reports Appeal Cases Series |
| All ER | All England Law Reports |
| BLR | Building Law Reports |
| CILL | Construction Industry Law Letter |
| CL | Construction Law |
| CLD | Construction Law Digest |
| CLY | Construction Law Yearbook |
| Const. LJ | Construction Law Journal |
| EG | Estates Gazette |
| LR | Law Reports First Series |
| Lloyd's Rep | Lloyd's List Law Reports |
| NSWLR | New South Wales Law Reports |

Appleby v. Myers [1867] LR 2 CP 651 .............................. 12.1
Architectural Installation Services Ltd v. James Gibbons Windows Ltd (1989)
    46 BLR 91 ................................................ 15.1
Aughton Ltd v. MF Kent Services Ltd (1991) 57 BLR 1 ................. 4.3
Ben Barrett & Son (Brickwork) Ltd v. Henry Boot Management Ltd (1995)
    CILL 1026 ................................................ 4.3
Bovis Construction (Scotland) Ltd v. Whatlings Construction Ltd (1994)
    67 BLR 25 ................................................ 15.1
Bramall & Ogden v. Sheffield City Council (1983) 29 BLR 73 ........... 3.11
Cape Durasteel Ltd v. Rosser and Russell Building Services Ltd (1995)
    CL Vol. 6 Issue 5 ......................................... 14.2
Channel Tunnel Group Ltd v. Balfour Beatty Construction Ltd (1993)
    61 BLR 1 ................................................. 14.8
Davis Contractors v. Fareham UDC [1956] AC 696 .................... 6.11
Emson Eastern Ltd v. EME Developments Ltd (1991) 55 BLR 114 ........ 6.3
Greater London Council v. Cleveland Bridge & Engineering Co. Ltd (1986)
    34 BLR 50 ................................................ 8.1
Gleeson (MJ) Group plc v. Wyatt of Snetterton Ltd (1994) 72 BLR 15 .... 14.5
Glenlion Construction Ltd v. The Guinness Trust (1987) 39 BLR 89 ...... 8.5
Greaves (Contractors) Ltd v. Baynham Meikle & Partners (1975) 4 BLR 56 ... 7.2
Hancock v. Brazier [1966] 2 All ER 901 ............................ 11.1

Heyman v. Darwins [1942] AC 356 ............................... 15.1
Hill (JM) & Sons Ltd v. Camden London Borough Council (1980) 18 BLR 31
................................................................ 15.1
Humber Oil Trustees Ltd v. Harbour and General Works (Stevin) Ltd (1991)
   59 BLR 1 ................................................... 11.2
Independent Broadcasting Authority v. EMI Electronics (1980) 14 BLR 1
................................................................ 3.7, 7.2
Langham House Developments Ltd v. Brompton Securities Ltd (1980)
   265 Estates Gazette 719 ..................................... 14.2
Lockland Builders Ltd v. John Kim Rickwood (1995) 13 CLD 07 31 ...... 15.1
Lodder v. Slowey [1904] AC 442 73 ............................... 15.3
London Borough of Merton v. Stanley Hugh Leach Ltd (1985) 32 BLR 51 .. 5.3
Lubenham Fidelities & Investments Co. Ltd v. South Pembrokeshire
   District Council (1986) 33 BLR 39 ............................ 15.1
Milburn Services Ltd v. United Trading Group (UK) Ltd (1995) CILL 1109
................................................................ 11.1
National Trust for Places of Historic Interest or Natural Beauty (The) v.
   Haden Young Ltd (1993) 66 BLR 88 ........................... 13.1
Nevill (HW) (Sunblest) Ltd v. William Press & Sons (1981) 20 BLR 78 ... 6.3, 8.6
Pigott Foundations Ltd v. Shepherd Construction Ltd (1993) 67 BLR 48 ... 16.4
Renard Constructions (ME) Pty. Ltd v. Minister of Public Works (1992)
   26 NSWLR 234 ............................................ 5.3, 15.3
Ruxley Electronics and Construction Ltd v. Forsyth (1995) 73 BLR 1 ... 9.7, 9.8
Shanks & McEwan (Contractors) Ltd v. Strathclyde Regional Council [1994]
   CILL 916 .................................................. 7.4
Surrey Heath Borough Council v. Lovell Construction (1988) 42 BLR 25 ... 8.6
Temloc Ltd v. Errill Properties Ltd (1987) 39 BLR 30 ............... 3.6, 3.11
Thorn v. London Corporation (1876) 1 AC ......................... 6.11
Trafalgar House Construction (Regions) Ltd v. General Surety & Guarantee Co.
   Ltd (1995) 73 BLR 32 ....................................... 3.2
Turner and Sons Ltd v. Mathind Ltd (1986) 5 Const. LJ 273 ......... 3.6, 3.11
Turriff Ltd v. Welsh National Water Development Authority (1980)
   Construction Law Yearbook (1994) Page 122 ................... 6.11
Wates Construction (London) Ltd v. Franthom Property Ltd (1991)
   53 BLR 23 ................................................. 3.9
William Tomkinson & Sons Ltd v. The Parochial Church Council of
   St. Michael in the Hamlet (1990) 6 Const. LJ 319 ................ 13.1
Yorkshire Water Authority Ltd v. Sir Alfred McAlpine & Son (Northern) Ltd
   (1985) 32 BLR 114 ......................................... 6.3

# Index note

The second edition of the New Engineering Contract contains a comprehensive index of subjects referenced to clause numbers. In this book a full table of clause numbers with descriptions is referenced to chapter sections. The table is set out in the following pages.

Readers of this book who wish to have the benefit of a subject index will find it a straightforward matter to move from the subjects in the NEC index to the chapter sections in this book.

# Table of clause references

**Note:** references are to chapter sections.

**Core clauses**

*1 General*

| | | | |
|---|---|---|---|
| **Actions** | 10 | | |
| | 10.1 | Actions | 5.3, 6.2 |
| **Identified and defined terms** | 11 | | |
| | 11.1 | Terms | 4.4, 6.3 |
| | 11.2(1) | The parties | 5.1, 6.3 |
| | 11.2(2) | Others | 5.2, 6.3 |
| | 11.2(3) | The contract date | 4.5, 6.3 |
| | 11.2(4) | To provide the works | 6.3 |
| | 11.2(5) | Works information | 4.6, 6.3 |
| | 11.2(6) | Site information | 4.7, 6.3 |
| | 11.2(7) | The site | 6.3 |
| | 11.2(8) | The working areas | 6.3 |
| | 11.2(9) | Subcontractors | 5.1, 6.3 |
| | 11.2(10) | Plant and materials | 6.3 |
| | 11.2(11) | Equipment | 6.3 |
| | 11.2(12) | The completion date | 6.3 |
| | 11.2(13) | Completion | 6.3 |
| | 11.2(14) | The accepted programme | 6.3 |
| | 11.2(15) | Defects | 6.3, 9.2 |
| | 11.2(16) | The defects certificate | 6.3, 9.2 |
| | 11.2(17) | The fee | 6.3 |
| | 11.2(18) | Not used | |
| | 11.2(19) | Definition of the prices | 10.9, 10.10 |
| | 11.2(20) | Definition of the prices | 10.5, 10.7 |
| | 11.2(21) | Definition of the prices | 10.6, 10.8 |
| | 11.2(22) | Definition of the price for work done to date | 10.10 |
| | 11.2(23) | Definition of the price for work done to date | 10.7, 10.8, 10.9 |
| | 11.2(24) | The price for work done to date | 10.5 |
| | 11.2(25) | The price for work done to date | 10.6 |

## Table of clause references

| | 11.2(26) | Definition of actual cost .................... 10.10 |
| --- | --- | --- |
| | 11.2(27) | Definition of actual cost ........... 10.7, 10.8, 10.9 |
| | 11.2(28) | Definition of actual cost ............... 10.5, 10.6 |
| | 11.2(29) | Definition of disallowed cost .............. 10.10 |
| | 11.2(30) | Definition of disallowed cost ....... 10.7, 10.8, 10.9 |

**Interpretation and the law**    **12**
- 12.1    Interpretation ............................. 6.4
- 12.2    Law of the contract ....................... 6.4

**Communications**    **13**
- 13.1    Communications ......................... 5.10
- 13.2    Receipt of communications ................ 5.10
- 13.3    Period for reply .......................... 5.10
- 13.4    Replies on acceptances .................... 5.10
- 13.5    Extending the period for reply ............. 5.10
- 13.6    Issue of certificates ....................... 5.10
- 13.7    Notifications ............................. 5.10
- 13.8    Withholding an acceptance ................ 5.10

**The project manager and the supervisor**    **14**
- 14.1    Acceptance of a communication ............ 5.11
- 14.2    Delegation .............................. 5.11
- 14.3    Instructions ............................. 5.11
- 14.4    Replacements ............................ 5.11

**Adding to the working areas**    **15**
- 15.1    Adding to the working areas ............... 6.7

**Early warning**    **16**
- 16.1    Early warning notices ..................... 6.8
- 16.2    Attendance at early warning meetings ........ 6.8
- 16.3    Early warning meetings ................... 6.8
- 16.4    Records of early warning meetings ........... 6.8

**Ambiguities and inconsistencies**    **17**
- 17.1    Ambiguities and inconsistencies ............ 4.10

**Health and safety**    **18**
- 18.1    Health and safety ......................... 6.10

**Illegal and impossible requirements**    **19**
- 19.1    Illegal and impossible requirements ......... 6.11

## 2 The contractor's main responsibilities

| | | | |
|---|---|---|---|
| **Providing the works** | **20** | | |
| | 20.1 | Obligation to provide the works | 7.3 |
| | 20.2 | Management obligations | 7.3 |
| | 20.3 | Practical implications of design and subcontracting | 7.3 |
| | 20.4 | Forecasts of total actual costs | 7.3 |
| **The contractor's design** | **21** | | |
| | 21.1 | The contractor's design | 7.4 |
| | 21.2 | Acceptance of the contractor's design | 7.4 |
| | 21.3 | Submission of design in parts | 7.4 |
| | 21.4 | Indemnity against claims | 7.4 |
| | 21.5 | Limitation of liability | 7.4 |
| **Using the contractor's design** | **22** | | |
| | 22.1 | Employer's use of the contractor's design | 7.5 |
| **Design of equipment** | **23** | | |
| | 23.1 | Design of equipment | 7.6 |
| **People** | **24** | | |
| | 24.1 | Key persons | 7.7 |
| | 24.2 | Removal of an employee | 7.7 |
| **Co-operation** | **25** | | |
| | 25.1 | Co-operation with others | 7.8 |
| **Subcontracting** | **26** | | |
| | 26.1 | Responsibility for subcontractors | 7.9 |
| | 26.2 | Acceptance of subcontractors | 7.9 |
| | 26.3 | Conditions of subcontracts | 7.9 |
| | 26.4 | Contract data for subcontracts | 7.9 |
| **Approval from others** | **27** | | |
| | 27.1 | Approval from others | 7.10 |
| **Access to work** | **28** | | |
| | 28.1 | Access to work | 7.11 |
| **Instructions** | **29** | | |
| | 29.1 | Instructions | 7.12 |

## 3 Time

**Starting and completion**   **30**
- 30.1   Starting and completion .................... 8.2
- 30.2   Deciding and certifying completion .......... 8.2

**The programme**   **31**
- 31.1   Submission of the programme .............. 8.3
- 31.2   Detail of the programme ................... 8.3
- 31.3   Acceptance of programmes ................. 8.3
- 31.4   Activities in the programme ................ 8.3

**Revising the programme**   **32**
- 32.1   Revised programmes ...................... 8.4
- 32.2   Submission of revised programmes .......... 8.4

**Possession of the Site**   **33**
- 33.1   Possession ............................... 8.6
- 33.2   Access, facilities and services .............. 8.6

**Instructions to stop or not to start work**   **34**
- 34.1   Instructions to stop or restart work........... 8.7

**Take over**   **35**
- 35.1   Take over and possession .................. 8.8
- 35.2   Take over and completion .................. 8.8
- 35.3   Take over and use of the works ............. 8.8
- 35.4   Certifying take over ....................... 8.8

**Acceleration**   **36**
- 36.1   Acceleration ............................. 8.9
- 36.2   Quotation for acceleration .................. 8.9
- 36.3   Accepting a quotation (A, B, C, D) ........... 8.9
- 36.4   Accepting a quotation (E, F) ................ 8.9
- 36.5   Subcontractors (C, D, E, F) ................. 8.9

## 4 Testing and defects

**Tests and inspections**   **40**
- 40.1   Tests and inspections...................... 9.3
- 40.2   Materials, facilities and samples ............. 9.3
- 40.3   Notifications.............................. 9.3
- 40.4   Repeat tests and inspections ................ 9.3
- 40.5   The supervisor's tests and inspections ........ 9.3
- 40.6   Costs of repeat tests and inspections ......... 9.3

**Testing and inspection before delivery**    **41**
- 41.1    Delivery of plant and materials .............. 9.4

**Searching and notifying defects**    **42**
- 42.1    Instructions to search ....................... 9.5
- 42.2    Notification of defects ...................... 9.5

**Correcting defects**    **43**
- 43.1    Correcting defects .......................... 9.6
- 43.2    Issue of the defects certificate .............. 9.6
- 43.3    Access for correcting defects ................ 9.6

**Accepting defects**    **44**
- 44.1    Proposals to accept defects ................. 9.7
- 44.2    Acceptance of defects ...................... 9.7

**Uncorrected defects**    **45**
- 45.1    Uncorrected defects ........................ 9.8

## 5 Payment

**Assessing the amount due**    **50**
- 50.1    Assessment procedure ..................... 10.2
- 50.2    The amount due ........................... 10.2
- 50.3    Failure to submit programme .............. 10.2
- 50.4    Application by the contractor .............. 10.2
- 50.5    Corrections of assessments ................ 10.2
- 50.6    Assessing the amount due ................ 10.7
- 50.7    Assessing the amount due ................ 10.9

**Payment**    **51**
- 51.1    Certification and obligations ............... 10.3
- 51.2    Time for payment and late payment ......... 10.3
- 51.3    Interest on corrected amounts .............. 10.3
- 51.4    Failure to certify ........................... 10.3
- 51.5    Rate of interest ............................ 10.3

**Actual cost**    **52, 53, 54**
- 52.1    Actual cost ................................ 10.4
- 52.2    Records of actual cost ..................... 10.7
- 52.3    Inspection of records ...................... 10.7
- 53.1    Calculating the share ...................... 10.7
- 53.2    Payment of the share ...................... 10.7
- 53.3    First assessment of the share .............. 10.7

*Table of clause references* 315

| | |
|---|---|
| 53.4 | Final assessment of the share .............. 10.7 |
| 53.5 | Proposals for reducing actual cost .......... 10.7 |
| 54.1 | Information in the activity schedule ..... 10.5, 10.7 |
| 54.2 | Changes to activity schedule ........... 10.5, 10.7 |
| 54.3 | Reasons for not accepting a revised activity schedule ......................... 10.5, 10.7 |
| 55.1 | The bill of quantities ..................... 10.6 |

## 6 Compensation events
**Compensation events**  60

| | | |
|---|---|---|
| 60.1 | Compensation events ................ 11.1, 11.2 |
| 60.1(1) | Changes to the works information .......... 11.2 |
| 60.1(2) | Failure to give possession ................. 11.2 |
| 60.1(3) | Failure to provide something .............. 11.2 |
| 60.1(4) | Instructions to stop or not to start work ...... 11.2 |
| 60.1(5) | Failure to work within times ............... 11.2 |
| 60.1(6) | Failure to reply to a communication ......... 11.2 |
| 60.1(7) | Objects of interest ....................... 11.2 |
| 60.1(8) | Changing a decision ..................... 11.2 |
| 60.1(9) | Withholding an acceptance ................ 11.2 |
| 60.1(10) | Searches for defects ..................... 11.2 |
| 60.1(11) | Tests or inspections causing delay .......... 11.2 |
| 60.1(12) | Physical conditions ...................... 11.2 |
| 60.1(13) | Weather conditions ...................... 11.2 |
| 60.1(14) | Employer's risks ........................ 11.2 |
| 60.1(15) | Take over before completion ............... 11.2 |
| 60.1(16) | Failure to provide materials etc. ............ 11.2 |
| 60.1(17) | Correction of an assumption ............... 11.2 |
| 60.1(18) | Breach of contract by the employer ......... 11.2 |
| 60.2 | Judging physical conditions ............... 11.2 |
| 60.3 | Inconsistency in site information ........... 11.2 |
| 60.4 | Differences in quantities .................. 11.2 |
| 60.5 | Increased quantities ..................... 11.2 |
| 60.6 | Mistakes in bill of quantities ............... 11.2 |

**Notifying compensation events**  61

| | | |
|---|---|---|
| 61.1 | Notifications by the project manager ........ 11.3 |
| 61.2 | Quotations for proposed instructions ........ 11.3 |
| 61.3 | Notifications by the contractor ............. 11.3 |
| 61.4 | The project manager's decisions ............ 11.3 |
| 61.5 | Failure to give early warning .............. 11.3 |
| 61.6 | Assumptions on effects ................... 11.3 |
| 61.7 | No notification after the defects date ........ 11.3 |

# Table of clause references

| | | | |
|---|---|---|---|
| **Quotations for compensation events** | **62** | | |
| | 62.1 | Instructions for alternative quotations | 11.4 |
| | 62.2 | Quotations for compensation events | 11.4 |
| | 62.3 | Submission of quotations | 11.4 |
| | 62.4 | Instruction to submit a revised quotation | 11.4 |
| | 62.5 | Extending the time for quotations | 11.4 |
| **Assessing compensation events** | **63** | | |
| | 63.1 | Changes to prices | 11.5 |
| | 63.2 | Reductions in prices | 11.5 |
| | 63.3 | Delay to completion | 11.5 |
| | 63.4 | Failure to give early warning | 11.5 |
| | 63.5 | Time and risk allowances | 11.5 |
| | 63.6 | Assumptions on reactions | 11.5 |
| | 63.7 | Ambiguity or inconsistency | 11.5 |
| | 63.8 | Changes to the activity schedule | 11.5 |
| | 63.9 | Changes to the bill of quantities | 11.5 |
| | 63.10 | Subcontracted work | 11.5 |
| | 63.11 | Use of the shorter schedules | 11.5 |
| **The project manager's assessments** | **64** | | |
| | 64.1 | The project manager's assessment | 11.6 |
| | 64.2 | Assessed programme | 11.6 |
| | 64.3 | Notification of the project manager's assessments | 11.6 |
| **Implementing compensation events** | **65** | | |
| | 65.1 | Implementing compensation events | 11.7 |
| | 65.2 | No revision for later information | 11.7 |
| | 65.3 | Changes to forecast amounts | 11.7 |
| | 65.4 | Changes to prices and the completion date | 11.7 |
| | 65.5 | Subcontract compensation events | 11.7 |
| **7 Title** | | | |
| **The employer's title to equipment, plant and materials** | **70** | | |
| | 70.1 | Title to equipment, plant and materials | 12.2 |
| | 70.2 | Title within the working areas | 12.2 |

| | | | |
|---|---|---|---|
| **Marking equipment, plant and materials outside the working areas** | **71** | | |
| | 71.1 | Marking equipment .......................... | 12.3 |
| **Removing equipment** | **72** | | |
| | 72.1 | Removal of equipment ..................... | 12.4 |
| **Objects and materials within the site** | **73** | | |
| | 73.1 | Articles of interest ........................ | 12.5 |
| | 73.2 | Title to materials ......................... | 12.5 |

## 8 Risks and insurances

| | | | |
|---|---|---|---|
| **Employer's risks** | **80** | | |
| | 80.1 | Employer's risks .......................... | 13.2 |
| **The contractor's risks** | **81** | | |
| | 81.1 | The contractor's risks...................... | 13.3 |
| **Repairs** | **82** | | |
| | 82.1 | Repairs .................................. | 13.4 |
| **Indemnity** | **83** | | |
| | 83.1 | Indemnity................................ | 13.5 |
| | 83.2 | Contributory reduction ..................... | 13.5 |
| **Insurance cover** | **84** | | |
| | 84.1 | Provision of insurances .................... | 13.6 |
| | 84.2 | The insurance table ....................... | 13.6 |
| **Insurance policies** | **85** | | |
| | 85.1 | Submission of policies ..................... | 13.7 |
| | 85.2 | Waiver of subrogation ..................... | 13.7 |
| | 85.3 | Compliance with policies .................. | 13.7 |
| | 85.4 | Unrecovered amounts ..................... | 13.7 |
| **If the contractor does not insure** | **86** | | |
| | 86.1 | Contractor's failure to insure .............. | 13.8 |
| **Insurance by the employer** | **87** | | |
| | 87.1 | Submission of policies ..................... | 13.9 |
| | 87.2 | Acceptance of policies ..................... | 13.9 |
| | 87.3 | Failure by the employer to insure ........... | 13.9 |

## 9 Disputes and termination

**Settlement of disputes** — **90**
- 90.1 Submission to adjudication ... 14.5
- 90.2 The adjudicator's decision ... 14.5

**The adjudication** — **91**
- 91.1 The timescale of the adjudication ... 14.6
- 91.2 Joinder of subcontract disputes ... 14.6

**The adjudicator** — **92**
- 92.1 The adjudicator ... 14.7
- 92.2 Appointment of a new adjudicator ... 14.7

**Review by the tribunal** — **93**
- 93.1 Reference to the tribunal ... 14.8
- 93.2 Powers of the tribunal ... 14.8

**Termination** — **94**
- 94.1 Notification of termination ... 15.4
- 94.2 The termination table ... 15.4
- 94.3 Implementation of termination procedures ... 15.4
- 94.4 Certification of amount due on termination ... 15.4
- 94.5 Cessation on termination ... 15.4

**Reasons for termination** — **95**
- 95.1 Insolvency ... 15.5
- 95.2 Contractor's defaults ... 15.5
- 95.3 Contractor's continuing defaults ... 15.5
- 95.4 Failure to pay ... 15.5
- 95.5 Frustration ... 15.5
- 95.6 Prolonged suspension ... 15.5

**Procedures on termination** — **96**
- 96.1 Completion of the works ... 15.6
- 96.2 Withdrawal from the site ... 15.6

**Payment on termination** — **97**
- 97.1 Valuation at termination ... 15.7
- 97.2 Other amounts due ... 15.7
- 97.3 Payment on termination – Option A ... 15.7
- 97.4 Payment on termination – Options C and D ... 15.7

*Table of clause references*

## Main option clauses

### Option A: priced contract with activity schedule

| | | | |
|---|---|---|---|
| **Identified and defined terms** | **11** | | |
| | 11.2(20) | Definition of the prices | 10.5, 10.7 |
| | 11.2(24) | The price for work done to date | 10.5 |
| | 11.2(28) | Definition of actual cost | 10.5, 10.6 |
| **The programme** | **31** | | |
| | 31.4 | Activities in the programme | 8.3 |
| **Acceleration** | **36** | | |
| | 36.3 | Accepting a quotation (A, B, C, D) | 8.9 |
| **The activity schedule** | **54** | | |
| | 54.1 | Information in the activity schedule | 10.5, 10.7 |
| | 54.2 | Changes to the activity schedule | 10.5, 10.7 |
| | 54.3 | Reasons for not accepting a revised activity schedule | 10.5, 10.7 |
| **Assessing compensation events** | **63** | | |
| | 63.8 | Changes to the activity schedule | 11.5 |
| | 63.10 | Subcontract work | 11.5 |
| | 63.11 | Use of the shorter schedules | 11.5 |
| **Implementing compensation events** | **65** | | |
| | 65.4 | Changes to prices and completion date | 11.7 |
| **Payment on termination** | **97** | | |
| | 97.3 | Payment on termination – Option A | 15.7 |

### Option B: Priced contract with bill of quantities

| | | | |
|---|---|---|---|
| **Identified and defined terms** | **11** | | |
| | 11.2(21) | Definition of the prices | 10.6 |
| | 11.2(25) | The price for work done to date | 10.6 |
| | 11.2(28) | Definition of actual cost | 10.5, 10.6 |
| **Acceleration** | **36** | | |
| | 36.3 | Accepting a quotation (A, B, C, D) | 8.9 |
| **The bill of quantities** | **55** | | |
| | 55.1 | The bill of quantities | 10.6 |

| | | |
|---|---|---|
| **Compensation events** | **60** | |
| | 60.4 | Differences in quantities .................. 11.2 |
| | 60.5 | Increased quantities ...................... 11.2 |
| | 60.6 | Mistakes in bill of quantities .............. 11.2 |
| **Assessing compensation events** | **63** | |
| | 63.9 | Changes to the bill of quantities ............ 11.5 |
| | 63.10 | Subcontracted work ....................... 11.5 |
| | 63.11 | Use of the shorter schedule ................ 11.5 |
| **Implementing compensation events** | **65** | |
| | 65.4 | Changes to prices and the completion date ... 11.7 |

*Option C: Target contract with activity schedule*

| | | |
|---|---|---|
| **Identified and defined terms** | **11** | |
| | 11.2(20) | Definition of the prices ............... 10.5, 10.7 |
| | 11.2(23) | Definition of the price for work done to date .. 10.7 |
| | 11.2(27) | Definition of actual cost................... 10.7 |
| | 11.2(30) | Disallowed cost ......................... 10.7 |
| **Providing the works** | **20** | |
| | 20.3 | Practical implications of design and subcontracting ........................................ 7.3 |
| | 20.4 | Forecasts of total actual costs .............. 7.3 |
| **Subcontracting** | **26** | |
| | 26.4 | Contract data for subcontracts ............. 7.9 |
| **The programme** | **31** | |
| | 31.4 | Activities in the programme ................ 8.3 |
| **Acceleration** | **36** | |
| | 36.3 | Accepting a quotation (A, B, C, D) ........... 8.9 |
| | 36.5 | Subcontractors (C, D, E, F) ................. 8.9 |
| **Assessing the amount due** | **50** | |
| | 50.6 | Assessing the amount due ................ 10.7 |
| **Actual cost** | **52** | |
| | 52.2 | Records of actual cost .................... 10.7 |
| | 52.3 | Inspection of records ..................... 10.7 |

## Table of clause references

| | | | |
|---|---|---|---|
| The contractor's share | 53 | | |
| | 53.1 | Calculating the share | 10.7 |
| | 53.2 | Payment of the share | 10.7 |
| | 53.3 | First assessment of the share | 10.7 |
| | 53.4 | Final assessment of the share | 10.7 |
| | 53.5 | Proposals for reducing actual cost | 10.7 |
| The activity schedule | 54 | | |
| | 54.1 | Information in the activity schedule | 10.5, 10.7 |
| | 54.2 | Changes to activity schedule | 10.5, 10.7 |
| | 54.3 | Reasons for not accepting a revised activity schedule | 10.5, 10.7 |
| Assessing compensation events | 63 | | |
| | 63.8 | Changes to the activity schedule | 11.5 |
| | 63.11 | Use of the shorter schedule | 11.5 |
| Implementing compensation events | 65 | | |
| | 65.4 | Changes to prices and the completion date | 11.7 |
| Payment on termination | 97 | | |
| | 97.4 | Payment on termination – Options C and D | 15.7 |

### Option D: Target contract with bill of quantities

| | | | |
|---|---|---|---|
| Identified and defined terms | 11 | | |
| | 11.2(21) | Definition of the prices | 10.6 |
| | 11.2(23) | Definition of the price for work done to date | 10.7 |
| | 11.2(27) | Definition of actual cost | 10.7 |
| | 11.2(30) | Disallowed cost | 10.7 |
| Providing the works | 20 | | |
| | 20.3 | Practical implications of design and subcontracting | 7.3 |
| | 20.4 | Forecasts of total actual costs | 7.3 |
| Subcontracting | 26 | | |
| | 26.4 | Contract data for subcontractors | 7.9 |
| Acceleration | 36 | | |
| | 36.3 | Accepting a quotation (A, B, C, D) | 8.9 |
| | 36.5 | Subcontractors (C, D, E, F) | 8.9 |

| | | | |
|---|---|---|---|
| **Assessing the amount due** | 50 | | |
| | 50.6 | Assessing the amount due | 10.7 |
| **Actual cost** | 52 | | |
| | 52.2 | Records of actual cost | 10.7 |
| | 52.3 | Inspection of records | 10.7 |
| **The contractor's share** | 53 | | |
| | 53.1 | Calculating the share | 10.7 |
| | 53.2 | Payment of the share | 10.7 |
| | 53.3 | First assessment of the share | 10.7 |
| | 53.4 | Final assessment of the share | 10.7 |
| | 53.5 | Proposals for reducing actual cost | 10.7 |
| **Bill of quantities** | 55 | | |
| | 55.1 | The bill of quantities | 10.6 |
| **Compensation events** | 60 | | |
| | 60.4 | Differences in quantities | 11.2 |
| | 60.5 | Increased quantities | 11.2 |
| | 60.6 | Mistakes in bill of quantities | 11.2 |
| **Assessing compensation events** | 63 | | |
| | 63.8 | Changes to the bill of quantities | 11.5 |
| | 63.11 | Use of the shorter schedule | 11.5 |
| **Implementing compensation events** | 65 | | |
| | 65.4 | Changes to prices and the completion date | 11.7 |
| **Payment on termination** | 97 | | |
| | 97.4 | Payment on termination – Options C and D | 15.7 |

*Option E: Cost reimbursable contract*

| | | | |
|---|---|---|---|
| **Identified and defined terms** | 11 | | |
| | 11.2(19) | Definition of the prices | 10.9 |
| | 11.2(23) | Definition of the price for work done to date | 10.7 |
| | 11.2(27) | Definition of actual cost | 10.7 |
| | 11.2(30) | Disallowed cost | 10.7 |
| **Providing the works** | 20 | | |
| | 20.3 | Practical implications of design and subcontracting | 7.3 |

*Table of clause references* 323

|  | 20.4 | Forecasts of total actual costs .............. 7.3 |
|---|---|---|
| **Subcontracting** | **26** | |
|  | 26.4 | Contract data for subcontracts ............. 7.9 |
| **Acceleration** | **36** | |
|  | 36.4 | Accepting a quotation (E, F) ............... 8.9 |
|  | 36.5 | Subcontractors (C, D, E, F) ................ 8.9 |
| **Assessing the amount due** | **50** | |
|  | 50.7 | Assessing the amount due ................ 10.9 |
| **Actual cost** | **52** | |
|  | 52.2 | Records of actual cost .................... 10.7 |
|  | 52.3 | Inspection of records ..................... 10.7 |
| **Assessing compensation events** | **63** | |
|  | 63.11 | Use of the shorter schedule ............... 11.5 |
| **Implementing compensation events** | **65** | |
|  | 65.3 | Changes to forecast amounts .............. 11.7 |
|  | 65.5 | Subcontract compensation events .......... 11.7 |

## *Option F: Management contract*

| **Identified and defined terms** | **11** | |
|---|---|---|
|  | 11.2(19) | Definition of the prices ................... 10.9 |
|  | 11.2(22) | Definition of the price for work done to date . 10.10 |
|  | 11.2(26) | Definition of actual cost.................. 10.10 |
|  | 11.2(29) | Definition of disallowed cost.............. 10.10 |
| **Providing the works** | **20** | |
|  | 20.2 | Management obligations ................... 7.3 |
|  | 20.3 | Practical implications of design and subcontracting ........................... 7.3 |
|  | 20.4 | Forecasts of total actual costs .............. 7.3 |
| **Subcontracting** | **26** | |
|  | 26.4 | Contract data for subcontracts ............. 7.9 |
| **Acceleration** | **36** | |
|  | 36.4 | Accepting a quotation (E, F) ............... 8.9 |
|  | 36.5 | Subcontractors (C, D, E, F) ................ 8.9 |
| **Assessing the amount due** | **50** | |
|  | 50.7 | Assessing the amount due ................ 10.9 |

| | | | |
|---|---|---|---|
| **Actual cost** | 52 | | |
| | 52.2 | Records of actual cost | 10.7 |
| | 52.3 | Inspection of records | 10.7 |
| **Implementing compensation events** | 65 | | |
| | 65.3 | Changes to forecast amounts | 11.7 |
| | 65.5 | Subcontract compensation events | 11.7 |